AF572766

Pesticides in Food

PESTICIDES IN FOOD

ISSUES AND PERSPECTIVES

Edited by

Amrita Chakraborty

2010

Icfai Books
The Icfai University Press

PESTICIDES IN FOOD: ISSUES AND PERSPECTIVES

Editor: Amrita Chakraborty

First Edition: 2010
Printed in India

Published by

The Icfai University Press
52, Nagarjuna Hills, Punjagutta
Hyderabad, India – 500 082
Phone: (+91) (040) 23430–368, 369, 370, 372, 373, 374
Fax: (+91) (040) 23352521, 23435386
E-mail: info@icfaibooks.com, icfaibooks@icfai.org, ssd@icfai.org

ISBN: 9788131411698

Editorial Team: Divya Nigam and S Sreedar
Quality Support: M Manorama and J K Vadana

Contents

Section III

Remedial Measures

Overview

On one hand, agrochemicals are an essential requirement to boost farm productivity. They also ensure crop quality. As a consequence, we cannot wish away the use of agrochemicals in our food crops, fruits, vegetables etc. In a way, the entire food basket is also a basket of pesticides, because none of the mass consumption vegetables and fruits grown today whether in India or elsewhere in the world is without the extensive use of agrochemicals at some stage of production and/or shipment. In heavily populated, developing countries, where one has to coax the land to feed growing number of mouths, one has to constantly worry about enhancing land productivity and hence, the usage of agrochemicals is quite steep. Therefore, there is a trade-off between land productivity and risk from the amount of pesticide usage/residues in food crops/plants. And there is a need to be realistic, while being careful.

In recent years, there has been much hype around pesticide residues in soft drinks. Such shrill and excessive hype was uncalled for; after all, how often does the ordinary Indian have soft drinks? What he really has in large quantities are cereals, pulses, vegetables and fruits. How safe are they?

The Coke and Pepsi fiasco is an eye-opening saga in the agrochemicals-food safety debate. It is undoubtedly, easy to create a greater noise around highly visible, well known brands associated normally with the upper crust of society than with everyday products, from the many farms of unsung, unknown farmers. Nonetheless, what is the real picture of food safety in our entire food basket? Estimates show that in our country, the ADI (Acceptable Daily Intake) is exceeded by up to 7,000 percent[1] in some pesticides. Children are the most vulnerable and the worst affected. Their daily quota is exceeded manifold by different categories of food. And what is startling and eye-opening for an ordinary reader is the number of pesticides in fruits and vegetables and cereals. Have Coke and Pepsi been bashed up more because they are MNCs?

This book on agrochemicals in food and food safety gives us a highly credible peek into pesticides in cereals, pulses, vegetables and fruits. It is divided into 3 sections. They are Section I, 'An Introduction', Section II, 'Country Perspectives' and Section III, 'Remedial Measures'.

Section I: An Introduction

Article 1, **"The Benefits of Pesticides: A Story Worth Telling"**, a report by *Fred Whitford, David Pike, Greg Hanger, Frank Burroughs, Bill Johnson* and *Arlene Blessing,* on Pesticides named Purdue Pesticide Programs (PPP:70), reminds us of the use of pesticides in generating gains over risk. For decades, discussions among scientists and the public have focused on the real, predicted and perceived risks that pesticides pose to people and the environment. Overall, the public overlooks the benefits of pesticides. We have grown accustomed to buying fresh produce that is free of blemishes, and we expect canned fruits and vegetables to be free of insects. We cry wolf only when we hear the danger of pesticides residues. Hence, each one of us, whether scientists or regulators or government officials or the public/community, needs

---

[1] Down to Earth, August 25th 2006.

a realistic understanding of the risks associated with pesticide use. Pesticides need not be such a bad thing, since, in some cases, pesticides are the only viable alternative when there are significant risks associated with leaving certain pests uncontrolled. This article opens our eyes to the benefits of the much hated pesticide. Despite all these benefits it is difficult to convey to the public a sense of balance among pesticide benefits and risks. They highlight the risks and even equate it to poison (albeit at the prodding of some NGOs and their reports).

The next article, **"Agricultural Chemicals – How Much Input is Required, How Much is too Much?"** is written by *Gerd Fleischer*. Pesticides enter plants through metabolic processes and are rarely eliminated unaltered via roots or leaves except when shed in dead tissues. The use of chemical pesticides is one of the most disputed issues in agricultural development. For some, modern technology including a package of synthetic fertilizers, chemical pesticides and high-yielding seed varieties are the key for sustainable development, providing enough food and agricultural raw material for the world's growing population. For others, chemicals are an unnecessary evil that damage human health and the environment. Agrochemicals contribute to increased agricultural productivity if their use is limited and targeted. But considerable damage may be the result when farmers and other users have limited knowledge or where they get incentives for excessive use. When using chemicals, positive impacts should clearly outweigh the negative ones. Unfortunately, many farmers in developing countries, especially the emerging economies strive for short-term maximum yields with the help of excessive chemical use instead of focusing on more sustainable practices.

The succeding article, **"ICRA Sector Analysis: Insecticides/ Pesticides"** is a report by *ICRA Information, Grading and Research Service,* which reminds us of the crying issue of acute food contamination through the use of pesticides, specially in developing countries. According to estimates, such contamination arises 13 times more often in developing countries than in developed countries. The reasons for higher pesticide residues/contamination in developing countries are many; pesticides are used indiscriminately here and the security instructions and necessary protection measures are not respected, insufficient training and information on the risk of pesticide use and lack of MRLs and other yardsticks measuring the risk of pesticide use. According to a study in India, residue of DDT and benzene hexachloride, both suspected carcinogens, were found in breast milk samples collected from mothers in Punjab.

What is noteworthy is that overall pesticide use in India over the last decade has remained constant or declined. The reduction is explained partly by greater efficiency of pesticide use as a result of improvements in pest management practices and technology and government policies aimed at both improving pest management practices, and in some cases targeting a reduction in pesticide use.

Section II: Country Perspectives

Section II gives us different country perspectives. The first article in this section **"Your Daily Poison: The Second UK Pesticide Exposure Report"** by *Alison Craig,* gives us an analysis of results from UK regulators and the latest European Commission data on pesticides. Excerpts like 'I have lived in this Hampshire village, of about 500 residents, for nearly thirty years. In the last ten to twelve years, I have noticed a dramatic increase in the incidence of cancer. I can think of 26 people in the village who have either been diagnosed with cancer, or who have died from it recently. There are numerous cases of breast cancer and bone cancer. I am concerned with the factors that may be causing some of these cancers. We live adjacent to a farm of a thousand acres and crops are regularly sprayed', indicate that levels of pesticide residues in food are increasing in the UK and are significant across Europe. This may be an example of Pesticide exposure through the surrounding environment but pesticide exposure through food is also worrisome, when the pesticide residues exceed MRL and such foods are consumed on an everyday basis in fairly large quantities as staple food.

According to the UK testing program 2004, many commonly consumed foods contain Pesticide Residues greater than MRL in UK and the rest of Europe. For instance, the following foods have repeatedly been found with pesticide residues greater than MRL and, consequently, they are most likely to be potentially carcinogenic when consumed in fairly large quantities:

1. Apples, Captan, Suspected carcinogen USEPA B2; EU 3; IARC 3
2. Beans, Dicofol, Suspected carcinogen USEPA C; IARC 3
3. Cabbage, Iprodione, Suspected carcinogen USEPA L2; EU 3
4. Fish, Chlordane, Suspected carcinogen USEPA B2; EU 3;
5. Grapes, Carbaryl, Suspected carcinogen USEPA 2; EU 3; IARC 3.

The next article **"Pesticide Use and Pesticide Residues"** (excerpted from the report 'The Minimisation of Pesticide Residues in Food: A Review of the Published Literature') written by *D Atkinson, F Burnett, G N Foster, A Litterick, M Mullay* and *C A Watson.* The published information on pesticide residues indicates that a significant proportion of most major vegetable crops and all major fruit products consumed in the US (& Europe) contain residues of pesticides. Pesticide residues detected in samples analyzed show that commonly consumed fruits, vegetables and processed foods, considered as health foods (as apart from 'junk food') are not free of pesticide residues. A whole host of vegetables, from the ubiquitous potatoes to green beans etc to 'healthy' fruits like grapes, oranges, bananas and commonly consumed processed foods like Bread, Chocolate etc, all have pesticide residues. 82% of 251 blood samples in a study conducted in Belgium (Charlier & Plomteux 2002) were found to have detectable amounts of pesticide, in particular -p, p'- DDE (67% of samples). Urine was found to contain six pesticides or their metabolites in 50% of the samples of 978 adults participating in the Third National Health and Nutrition Examination in a study conducted in USA (Kieszak *et al.* 2002).

The next article **"The Rotten Apples: Experiences of Agrochemicals and Food Safety in India and Other Countries in the Developing World"** by *Amrita Chakraborty* reminds us that countries in the developed world, especially the US, implement food safety, which includes looking at dangerous amounts of pesticide residues in fruits and vegetables very stringently. The US has assigned clear responsibilities to two nodal agencies—the USEPA and the US Food and Drug Administration (USFDA). The former is the standard-setting agency. It is entrusted with registering pesticides for use. Before registration, it establishes ADI—what it calls Chronic Reference Dose—for that pesticide and sets MRLs (maximum residue limits/ tolerance limits) for residues in food commodities. It also makes sure that exposure is well below ADI.

The EU process is quite similar to what the US does. In global practices, the agency registering the pesticide establishes ADI, sets MRLs and then ensures that cumulative exposure is within the safety levels. In stark contrast in India, a pesticide is registered without any of these mandatory safety regulations. In fact, there is no legislative provision to link pesticide registration to setting MRLs. IA (Insecticide Act, 1968) mandates registration, but PFA, (the Prevention of Food Adulteration Act, 1954) mandates MRLs. Such legislative blindness has ensured that, of the 180 pesticides currently registered, MRLs have been set only for 71[2]. In other words, more than 60 percent of pesticides currently registered have no MRLs. Take sugar cane. The CIB [Central Insecticide Board] recommends 13 pesticides to be used. However, under PFA, MRLs for only 2 of the recommended pesticides have been established. Similarly in rice, 56 percent of recommended pesticides have no MRLs; in wheat, 43 percent have no MRLs; in mango, 44 percent. In coffee, 80 percent of recommended pesticides have no MRLs[3]. Against this background, the author wonders why has there been such a lot of brouhaha only around pesticides in soft drinks in India. The amount of unregistered pesticides in coffee (or mango, rice and all kinds of fruits and vegetables) in India almost

---

[2] Indian NGO Finds Pesticides in Colas, *http://www.corpwatch.org/article.php?id=9668.*

[3] Ibid.

makes consumption of soft drinks in large quantities a better alternative than freshly ground coffee!!

How many of us know that the extent of contamination of 'good food' like fruits and vegetables is dangerously high in India.

Section III: Remedial Measures

The first article in this section **"Recommendations for Future Action"** (excerpted from the report 'The Minimisation of Pesticide Residues in Food: A Review of the Published Literature') written by *D Atkinson, F Burnett, G N Foster, A Litterick, M Mullay* and *C A Watson,* is a review of publicly available information on the residues found in fresh foods currently being consumed in UK and of the scientific literature dealing with pesticide application. It says that pesticide residues in foods might be reduced by obviating the need for pesticide applications through the adoption of ecological methods of agriculture or by the use of modern biotechnological means of crop enhancement so that chemical crop protection is no longer required. In the shorter term, progress can be made through the continuation of conventional plant breeding, through modification of current systems of crop protection with chemicals and through changes in storage practices. Data from a range of sources indicate that much of the food currently consumed in the UK contains residues of pesticides. The remedy is that pesticide residues can be reduced through changes in farming practice. 'Recommendations for future action' gives a summary of potential actions to reduce selected residues in key crops.

The last but not the least important article of the book, **"Organic Farming: Its Relevance to the Indian Context"** by authors *P Ramesh, Mohan Singh* and *A Subba Rao,* talks of the increasing consciousness about health hazards associated with agrochemicals as well as conservation of environment. Consumers are showing a growing and clear preference for safe and hazard-free food, leading to the growing interest in alternate forms of agriculture in the world. There is no doubt that the trend is in favour of food with fewer pesticide residues because the demand for organic food is steadily increasing both in the developed and developing countries with an annual average growth rate of 20-25%.

However, there are certain pertinent issues that should be clarified before we go for a large-scale conversion to organic agriculture. Is the food produced by organic farming superior in quality? Is it economically feasible? In this article, these aspects of organic farming have been reviewed. The interest in organic agriculture in developing countries is growing because it requires less financial input and places more reliance on the natural and human resources available. Studies to date seem to indicate that organic agriculture offers comparative advantage in areas with less rainfall and relatively low natural and soil fertility levels. Organic agriculture does not need costly investments in irrigation, energy and external inputs, but rather organic agricultural policies have the potential to improve local food security, especially in marginal areas. However, a point to note is that large-scale conversion to organic agriculture would result in food shortage with the present state of knowledge and technology, as the yield reductions of organic systems relative to conventional agriculture average 10-15%, especially in intensive farming systems.

SECTION I

AN INTRODUCTION

1

The Benefits of Pesticides
A Story Worth Telling

Fred Whitford, David Pike, Greg Hanger, Frank Burroughs, Bill Johnson and Arlene Blessing

The use of pesticides has always remained a controversial topic of discussion among scientists and the public. There are significant risks associated with leaving certain pests uncontrolled; and, in some cases, pesticides are the only viable alternative. Properly used, pesticides provide benefits essential to the modern way of life. It is the need of the hour to have a realistic understanding of the risks associated with pesticide use. Uncontrolled pests can cause serious consequences to a person bitten by mosquitoes carrying West Nile virus and he may die, a child stung by bees, wasps, or ants may suffer a severe allergic reaction, a farmer's diseased tomatoes may be declined by the cannery, a load of wheat contaminated with wild garlic may be rejected by the mill, a homeowner may have to spend thousands of dollars to repair structural damage caused by termites.

Understanding Benefits Places Risks in Perspective

For decades, discussions among scientists and the public have focused on the real, predicted, and perceived risks that pesticides pose to people and the environment. Each use of a pesticide poses some level of risk, so it is not surprising that scientists, the regulated community, government officials, and the public need a realistic understanding of the risks associated with pesticide use. We must analyze how risk is assessed, identify the risks, and determine an appropriate level of concern.

There are significant risks associated with leaving certain pests uncontrolled; and, in some cases, pesticides are the only viable alternative. Properly used, pesticides provide benefits essential to our way of life. Uncontrolled pests can cause serious consequences:

- A person bitten by mosquitoes carrying West Nile virus may die.
- A child stung by bees, wasps, or ants may suffer a severe allergic reaction.
- A dog infested with fleas may become stressed to the point of illness.
- A farmer's diseased tomatoes may be declined by the cannery.
- A load of wheat contaminated with wild garlic may be rejected by the mill.
- A homeowner may have to spend thousands of dollars to repair structural damage caused by termites.

The benefits of pesticides commonly go unnoticed by the public. For example, if left unchecked, trees and brush growing beneath power lines would cause power outages. Herbicide use by utility companies to prevent undergrowth eliminates the problem and provides unobstructed access for maintenance and repairs. Road crews also use herbicides to control vegetation along highways, for safety reasons; clear roadsides increase visibility for drivers and allow water to escape more efficiently during a downpour or flooding. Herbicides also are used to fight invasive weeds in parks, wetlands, and natural areas.

Pesticides are used around our homes and businesses in ways we often take for granted. Plastics, paints, and caulks may contain fungicides to prevent mold. Toilet bowl cleaners and disinfectants often contain pesticides. Raw commodities and packaged grocery products—the foods we eat—are protected from insect contamination by the controlled use of insecticides in processing, manufacturing, and packaging facilities. Pesticides are used in grocery stores to manage insects and rodents attracted to food and food waste. There is little doubt that the proper use of pesticides improves our quality of life, protects our property, and promotes a better environment.

Benefits of Use Add to the Discussion of Risk

Understanding generates perspective, no matter the subject; and understanding the benefits of pesticides is essential to weighing the risks.

To identify potential risks associated with pesticide use, we must understand how risk is determined, what factors (including characteristics of the exposed population) control the potential for risk, what experience has shown about the risk, and what can be done to minimize the risk. Our conclusions then must be weighed against the benefits of pesticide use, factoring in any available alternatives as well as the benefits and risks of those alternatives.

The following discussion on driving provides a benefit versus risk analogy that can be used in our discussion of pesticides.

Statistically, driving motor vehicles is risky: every nine seconds a person is injured in an automobile collision, and every 12½ minutes someone dies; that is 42,000 deaths due to motor vehicle accidents every year in the United States. If we are injured, we are likely to suffer loss of income while recovering. If we damage public or private property, we are liable for repairing or replacing it. And if we are held liable for the injury or death of someone else, the financial consequences can be astronomical. If we die in an accident, our families suffer both emotionally and financially.

Motor vehicles also impact the environment. Our land and oceans are continually explored for oil in support of the vast quantities we use. Vehicle exhaust releases carbon monoxide into the air, and we expose ourselves to gasoline and diesel fuels, which are known carcinogens. We kill countless butterflies, birds, and mammals on the highways, which commonly are built on productive farmland and, sometimes, sensitive environmental areas. The risks and losses are real and measurable, but the benefits are obvious—and most of us consider motor vehicles essential to everyday life.

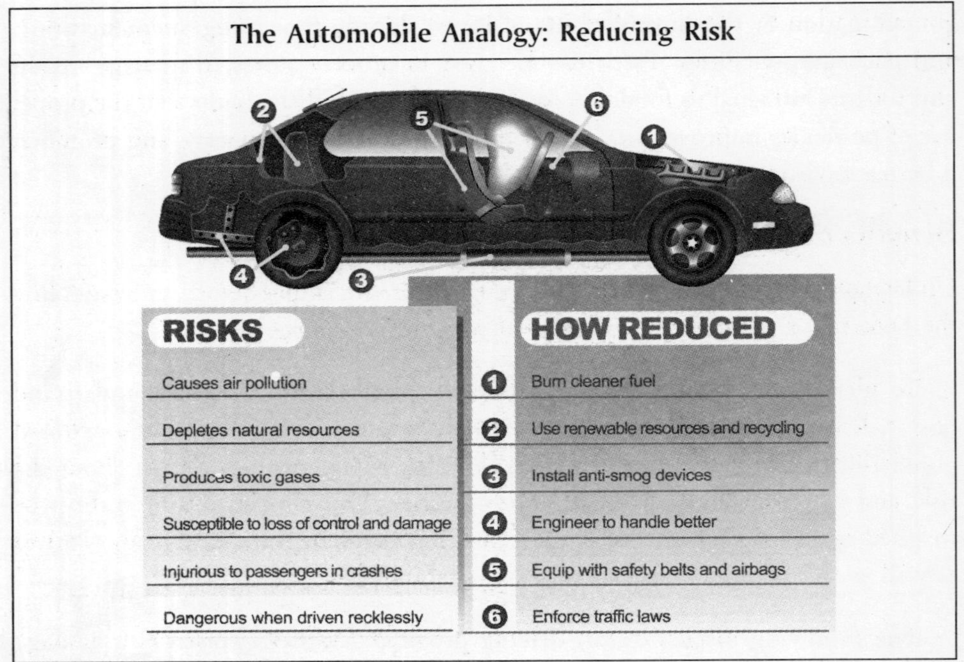

We generally agree that the benefits of driving a vehicle outweigh the risks of personal injury, death, or damage to the environment. But we are in control of certain decisions that affect the risk potential: what type of vehicle to purchase, how often it is used, how fast it is driven, whether to wear seat belts, etc. We continually seek to reduce driving risks by building safer vehicles and highways, by removing dangerous and incompetent drivers from the road, and by requiring drivers to pass a competency test and carry insurance. When we are in control of factors that contribute to risk, we accept it more readily.

The federal government regulates risks associated with modern technologies. For example, the Food and Drug Administration (FDA) reviews the benefits and risks associated with new medicines. FDA registration guidelines require pharmaceutical companies to write instructions for physicians and patients, indicating possible side effects and what actions to take if they occur. As a society, we clearly want the benefits of prescription and over-the-counter medications, but we expect associated risks to be minimized through regulatory oversight; and we have the benefit of a final, personal level of control in deciding whether or not to use them.

The public may not recognize or fully understand the benefits of pesticides, or they may take them for granted. Most people are not knowledgeable of the federal and state regulatory requirements and the extensive registration approval process that must be met before a pesticide can be offered for sale. Likewise, many are unaware of the training and testing that professional pesticide applicators and farmers undergo to become certified. Most states require certification and/or licensing of individuals who recommend or apply pesticides. Just as with medications prescribed by medical doctors, certain pesticides may be purchased only by licensed pesticide applicators and applied by or under the supervision of certified applicators. State and federal regulations govern how pesticide products may be used, and compliance is enforced at both levels.

Pesticide Benefits: A Story Worth Telling

Most of us acknowledge that the benefits of pesticides outweigh the risks, as evidenced by the continuing demand for more products to solve an ever increasing array of pest problems: bacteria in hospitals, mold in homes, the gypsy moth in hardwood forests, etc. The pesticide debate is somewhat analogous to that of the plastic industry. We benefit from the use of plastics in great quantities, but media stories and publicity generated a few years ago created uncertainty that the benefits

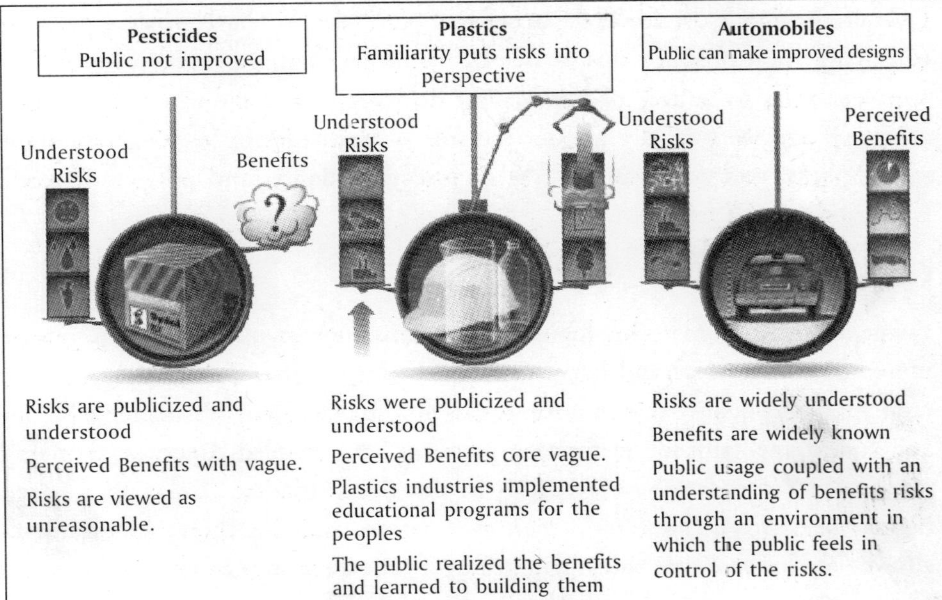

outweighed the risks. The plastics industry faced overwhelming opposition due to the risks: pollution of natural areas, landfill concerns, injury to wildlife (birds getting entangled in six-pack holders), human exposure to chemicals that might mimic human hormones, etc. But studies conducted by the American Plastics Council (*http://www.plastics.org*) suggested that providing the public factual information on the benefits of plastics would generate a more favorable attitude.

The American Plastics Council responded by running television commercials expounding the benefits of plastics in everyday products. The ads demonstrated that plastics are more than drinking cups and milk jugs. The council's message addressed the magnitude of plastics in our everyday lives: medicine containers, unbreakable plastic food containers, smoke alarms, eye glasses, sports and safety equipment, vehicle parts, refrigerators, soda containers, credit cards, toys, appliances, packaging materials, and much, much more. The advertising campaign made us realize that plastics provide benefits we take for granted. It also heightened environmental concerns over the volume of plastics entering our landfills, leading to widespread public participation in recycling programs.

Pesticides Provide Multiple Benefits

Overall, the public overlooks the benefits of pesticides. We have grown accustomed to buying fresh produce that is free of blemishes, and we expect canned fruits and vegetables to be free of insects. But do we give any thought to how this is achieved? Do we typically ponder the role of pesticides in keeping restaurants, malls, parks, and playgrounds free of insects, rodents, and poisonous weeds? It's a story worth telling.

Crop Production

Farmers benefit from technological innovations such as improved crop genetics, more efficient tractors and harvesters, and better fertilizers; they enjoy increased crop yield and quality due to advances in pest management. Pesticide technology and environmentally sound farming practices have enabled American farmers to manage large productive enterprises in the global market. Today they harvest higher yields from fewer acres than ever before, and this increased production allows the United States to contribute significant quantities of food when natural disasters occur around the world.

Pesticides play a key role in keeping malls (left) and parks (below) clean, safe, and attractive.

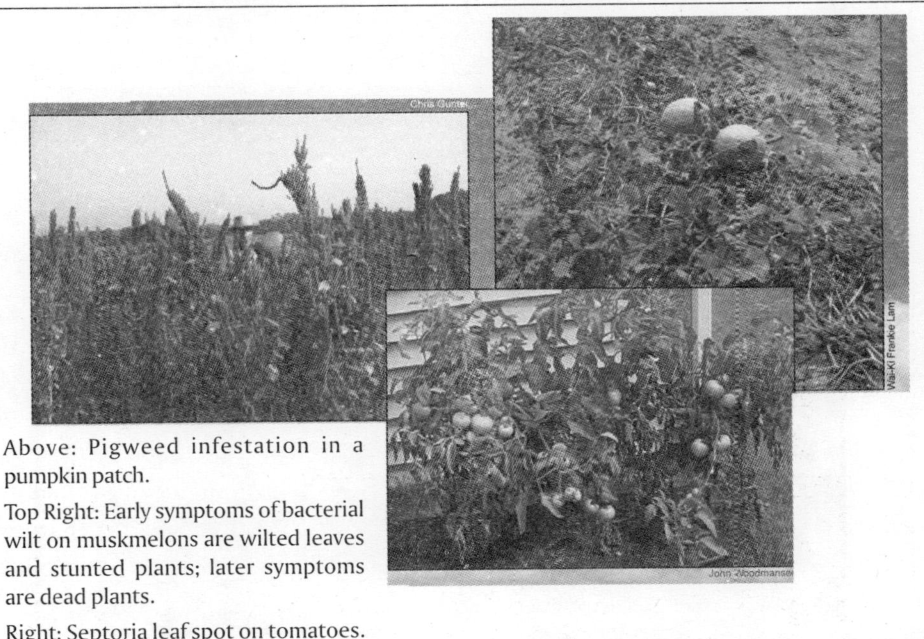

Above: Pigweed infestation in a pumpkin patch.

Top Right: Early symptoms of bacterial wilt on muskmelons are wilted leaves and stunted plants; later symptoms are dead plants.

Right: Septoria leaf spot on tomatoes.

The disease gummy stem blight has killed most of the green tissue of these watermelon vines; the melons will not ripen properly.

Benefits of Pesticides

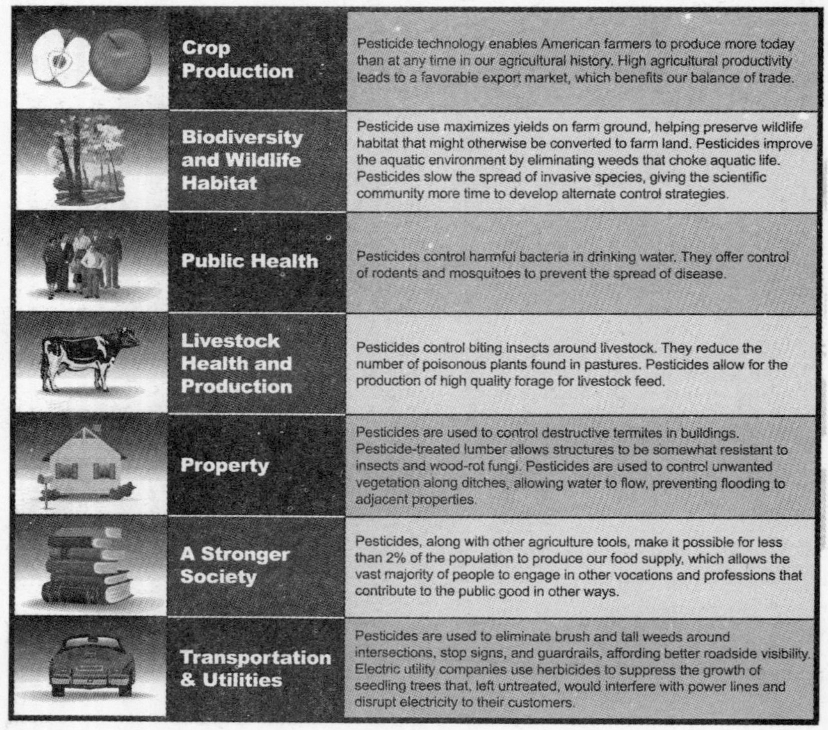

Crop Production	Pesticide technology enables American farmers to produce more today than at any time in our agricultural history. High agricultural productivity leads to a favorable export market, which benefits our balance of trade.
Biodiversity and Wildlife Habitat	Pesticide use maximizes yields on farm ground, helping preserve wildlife habitat that might otherwise be converted to farm land. Pesticides improve the aquatic environment by eliminating weeds that choke aquatic life. Pesticides slow the spread of invasive species, giving the scientific community more time to develop alternate control strategies.
Public Health	Pesticides control harmful bacteria in drinking water. They offer control of rodents and mosquitoes to prevent the spread of disease.
Livestock Health and Production	Pesticides control biting insects around livestock. They reduce the number of poisonous plants found in pastures. Pesticides allow for the production of high quality forage for livestock feed.
Property	Pesticides are used to control destructive termites in buildings. Pesticide-treated lumber allows structures to be somewhat resistant to insects and wood-rot fungi. Pesticides are used to control unwanted vegetation along ditches, allowing water to flow, preventing flooding to adjacent properties.
A Stronger Society	Pesticides, along with other agriculture tools, make it possible for less than 2% of the population to produce our food supply, which allows the vast majority of people to engage in other vocations and professions that contribute to the public good in other ways.
Transportation & Utilities	Pesticides are used to eliminate brush and tall weeds around intersections, stop signs, and guardrails, affording better roadside visibility. Electric utility companies use herbicides to suppress the growth of seedling trees that, left untreated, would interfere with power lines and disrupt electricity to their customers.

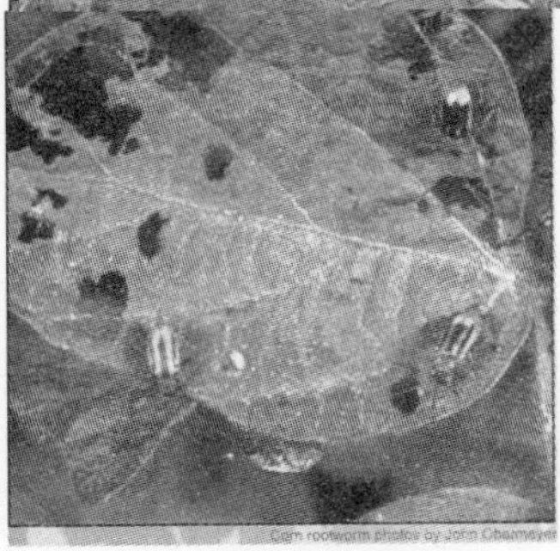

Corn rootworm photos by John Obermeyer

For decades, field crop producers in the Midwest were able to disrupt the corn rootworm life cycle using crop rotation. This prevented rootworm damage by larval feeding (top left and right) without the use of soil insecticides. Overtime, this cultural practice has altered the behavior of the adult beetle in some areas. Female beetles once laid eggs exclusively in corn, but now they will move to soybean and other non-corn crops to feed on foliage (far left) and lay eggs. Eggs laid in these fields overwinter in the soil and hatch the following spring where corn has been planted. The subsequent larval feeding causes lodging and "goose-necking" of corn plants (above) that significantly reduces yield and harvest ability.

Bill Johnson

A no-till soybean field is overtaken by golden ragwort (cressleaf groundsel). No-till farming would be impossible without herbicides for weed control.

Environment

Pesticides play a very important and prominent role in environmental protection.

Pesticides Promote Biodiversity

One very significant benefit of pesticide use is the preservation of wildlife habitat. Farmers who produce corn, milk, or vegetables make a living by the bushel, gallon, or pound; and they must produce enough to generate a decent profit. Pesticides help maximize profit by eliminating pests that reduce yields; the result is more product per acre, which lessens the need to convert natural areas such as woods and forests, native prairies, wetlands, plains, and other wildlife habitat into farm ground. Encouraging high levels of production on high-yielding land, thus preserving habitat critical to wildlife and native plant species, helps protect the planet's biodiversity.

Pesticides Contribute to Erosion Control

Generations of farmers tilled the land by hand or with animal- or tractor-powered implements, not only preparing the seed bed for planting but also controlling weeds and reducing insect and disease problems. These practices did little to control erosion, however, which became a major problem in farm fields that have steep slopes that allow soil to wash away during heavy rains.

Herbicides make no-till farming possible; they facilitate leaving residue from the previous crop in place to control soil erosion.

Our enjoyment of water recreation depends on the availability of clean water, and crop residue left in place during no-till farming—made possible by the availability of herbicides for weed control—helps block contaminants that might otherwise reach our waterways.

Over the last two decades, herbicides have made no-till farming a viable alternative, enabling farmers to reduce erosion by leaving the soil largely undisturbed. Herbicides are used for weed control in no-till crop production, eliminating the need for cultivation; residue from the previous crop holds the soil in place during wind and rain. Crop residue also impedes runoff of agricultural chemicals and soil that might otherwise affect aquatic habitat and fresh water supplies downstream.

Pesticides Control Invasive Pests

The acceleration of world trade has increased the potential for entrance of exotic pests into the United States. Once introduced, some pests aggressively replace native species, thus disturbing plant and wildlife ecology. Since invading species (e.g., Japanese beetle, gypsy moth, emerald ash borer, starlings) seldom bring along natural predators and diseases, their aggression toward crops, native plants, and human health goes unchallenged.

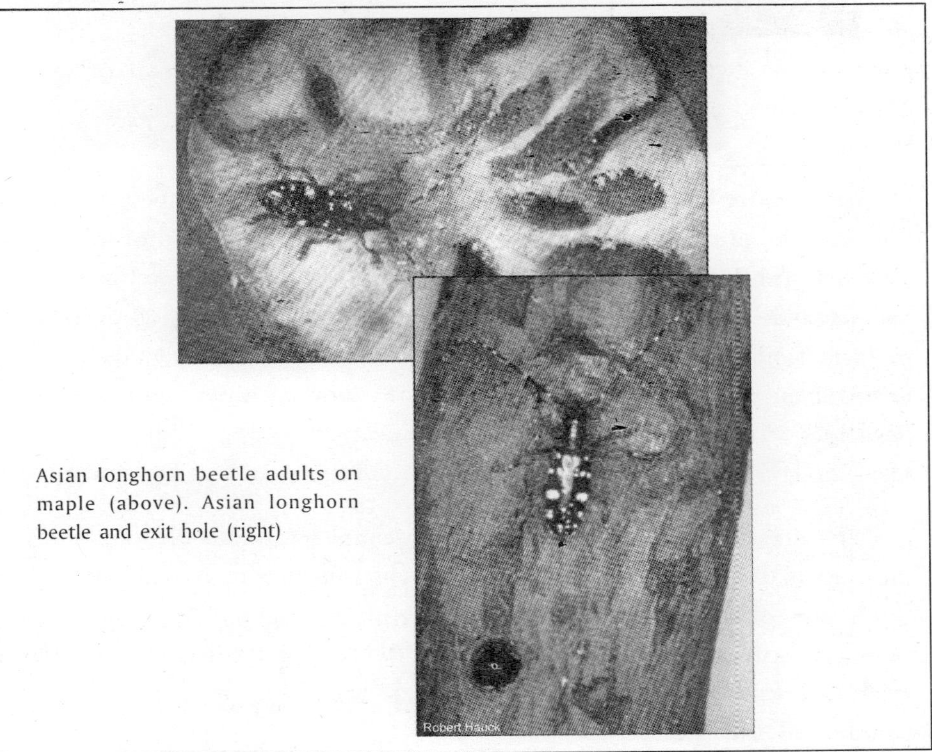

Asian longhorn beetle adults on maple (above). Asian longhorn beetle and exit hole (right)

Mature gypsy moth caterpillar on an oak leaf (below). Gyspy moth male and female (lower left). Defoliation of forest by gypsy moths (lower right).

Pesticides are commonly used to slow the spread of exotic pest populations. For example, piscicides have been useful in confining the spread of the northern snakehead fish in Maryland; herbicides control the weed purple loosestrife in marshes and wetlands, as well as Canada thistle and Johnsongrass along rights-of-way and in farm fields. Herbicides control aquatic weeds and help maintain a safe environment for boating, swimming, and recreational water sports nationwide. Pesticides often are used to contain new invasive species, buying time for the scientific community to develop alternative strategies such as biological control.

Left untreated, the Mediterranean fruit fly could devastate the fruit and vegetable industry in the United States; thus, annual monitoring of this insect is critical. Outbreaks of the Mediterranean fruit fly often originate from uncontrolled populations in residential fruit trees; and, when populations exceed threshold levels, areawide application of an insecticide is essential for the survival of the commercial fruit industry.

Dense garlic mustard rosettes carpeting a flood plain forest understory (upper left). Flowering garlic mustard (above). Closeup of garlic mustard, showing first-year leaves (left).

First adult emerald ash borer found in Indiana (Steuben Country) (left). Emerald ash borer larval damage underneath the bark of an ash free (below). Emerald ash borer larvae feeding beneath the bark of the tree (lower left).

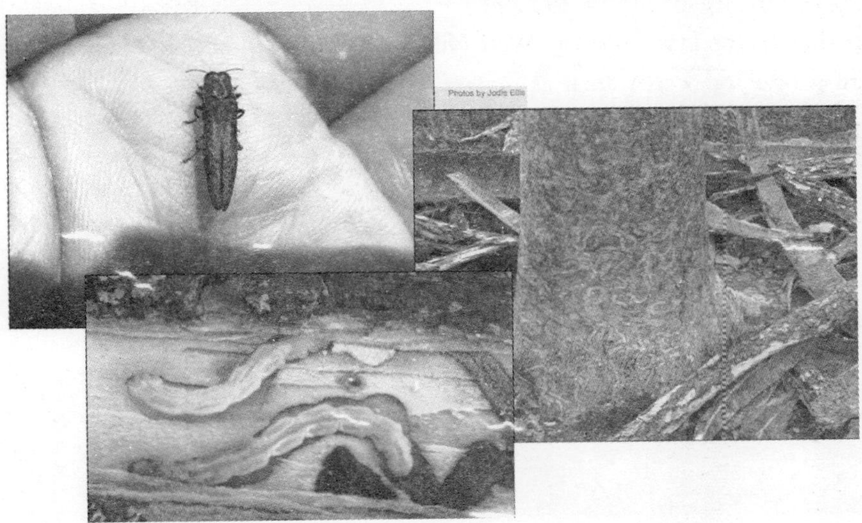

Health

Tom Campbell

Catherine Hill, a purdue entomology professor, balances a deer tick on her fingertip. Ticks can spread Lyme disease, Rocky Mountain spotted fever, and other diseases to humans and animals.

The role of pesticides in protecting public health is broad and varied. Water utility companies apply the pesticide chlorine to public drinking water to kill harmful bacteria. Pesticides known as disinfectants eliminate dangerous organisms that cause Legionnaire's disease, and hospitals rely on disinfectants to prevent the spread of bacteria such as *Staphylococcus.* Rodenticides are used in public housing units to control rodents that carry diseases such as the deadly hantavirus. Avicides are used to control birds near silos and grain storage buildings, reducing the likelihood of grain contamination and exposure of workers to the lung disease histoplasmosis. Healthcare professionals recommend the use of DEET to repel insects that vector Lyme disease, West Nile virus, malaria, etc. Herbicides control allergin-producing weeds such as ragweed and poison ivy.

Rosie Lerner

Poison ivy causes an itchy, blistery rash; severe cases sometimes warrant hospitalization

Livestock Production

Ranchers and farmers who raise cattle, sheep, goats, hogs, chickens, etc., are linked by a fundamental scientific principle: contented, well fed animals are better producers than their unattended counterparts.

Pesticides Control Infectious, Annoying Insects

Livestock is plagued by flies and other nuisance pests that vector disease, inflict painful bites, and cause stress. Insecticides control these insects when applied to the animals and/or their stalls, pens, corrals, barns, and containment facilities. As a result, the animals convert their feed into meat and milk more efficiently and advance to market quicker, thus lowering the cost of production. Consumers benefit directly from higher quality animal products.

Clemson University

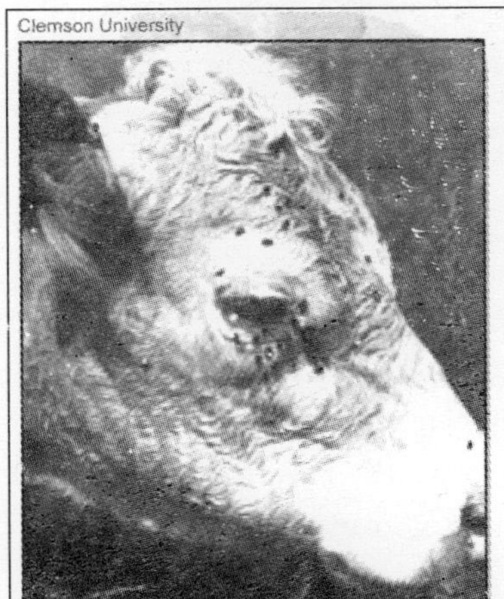

Face flies on a beef cow (left). Peggy K Powell, Ph.D., in a paper titled "Face Fly Biology and Management," West Virginia University Extension Service, 1995, stated that "12-14 flies per animal... can cause cattle to stop feeding and move into a shady location to escape from the flies, resulting in reduced animal production. Dairy cattle will cluster together to reduce face fly attack, thereby increasing heat stress and reducing milk production."

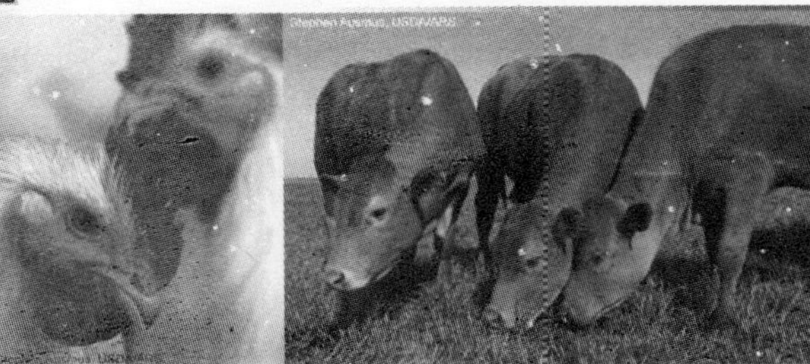

Pesticides Improve Feed Quality

Pesticides also play a key role in producing high quality grain and forage. Protein from alfalfa is an essential dietary requirement for dairy cattle, but alfalfa plants are very susceptible to diseases and insects. Infestations in alfalfa, left unchecked, can significantly reduce protein content and yield. Insecticides protect alfalfa crops from damaging insects, thereby contributing to our milk supply.

The alfalfa weevil larva (right) can come between a beautiful crop of alfalfa (above)

There are more than 50 noxious weeds that grow in forage crops and pastures in the United States; when eaten by livestock, they can cause symptoms ranging from sensitivity to the sun, to painful bleeding, to death. Livestock producers place high priority on preventing feed and range contamination from noxious weeds and seeds—and herbicides are one answer.

Insects also can be toxic, and their control is important to livestock producers. The presence of blister beetles in hay can kill horses or make them seriously ill. Insect feeding can damage plants and predispose them to fungal infections that can be poisonous to animals when concentrated in their feed.

Pesticides Improve Pastures

Horse, sheep, and cattle owners understand the importance of prudent herbicide use in maintaining productive pastures. Animals grow and stay healthy by converting grass and legumes into energy and muscle. Weeds crowd out quality plants on which the animals can graze; they are lower in protein and less palatable

Arlene Blessing

Herbicides were applied to kill noxious weeds in the pasture and woody brush in the fence row.

to animals and, over time, may spread and take over the pasture. When the quality or quantity of forage is insufficient, hay, grain, and feed must be purchased to supplement the animals' diet—and the cost is passed on to the consumer. Poisonous weeds can kill or injure horses, cows, sheep, and hogs. Herbicides are one of the answers.

Pesticides Keep Honey Bees Alive

Professional beekeepers contract with fruit and vegetable farmers to deliver truckloads of bee hives to their fields to facilitate the essential pollination of their crops. But Varoa mites that live as parasites on bees can significantly reduce honeybee populations. Fortunately, cloth strips treated with a miticide can be hung in hives to kill the mites. As bees enter and exit the hive, they pass through vapor from the strips, which kills the mites without harming the bees.

Property

Insecticides can be used to control termites, carpenter ants, and other structural insects. Museum directors use them to protect irreplaceable and extremely valuable collections from insect feeding; exhibits that contain plant, cloth, leather, and animal specimens are particularly vulnerable. The exhibit shown below is a second-century letter, written on papyrus (writing material made from the aquatic plant *Cyperus papyrus).* The letter's value is immeasurable, and it must be protected.

Pests can destroy hundreds or thousands of dollars' worth of plants in a very short time; and those with sentimental value (e.g., mother's old roses) cannot be replaced. Sometimes, pesticides are the best answer to protect the homeowner's investment and the plants she especially wants to maintain.

Exhibit

Courtesy of The William R. and Clarice V. Spurlock Museum, University of Illinois.

Oxyrhynchus Papyrus, No. 932 (1914.21.0010). A second-century letter, written in Greek, on papyrus, unearthed at the site of Oxyrhynchus in Egypt, first excavated between 1897 and 1907.

County drainage boards apply herbicides to ditch banks to eliminate trees and shrubs, thus keeping ditchwater flowing. Left untreated, woody vegetation shades the grass and causes thinning, which allows the release of soil into the ditches. As the ditches fill with soil, flooding increases on adjacent properties and can result in stagnant water that attracts mosquitoes.

The construction industry and homeowners use pesticide-treated lumber to build durable homes, decks, and docks resistant to insects and wood-rot fungi.

Aphids feeding on ornamental swamp milkweed (above). Hawthorn woolly aphids on hawthorn (right). Cottony maple scale (far right). A homeowner's beautiful landscaping—an investment worth protecting (above right).

Recreation

Pesticides also help maintain recreational areas. Golf courses and parks are more attractive due to the use of fungicides to combat turf diseases, insecticides to control destructive insects, and herbicides to control aggressive weeds. Herbicides are used on athletic fields to promote healthy turf, which has been shown to reduce sports injuries (fewer twisted knees and ankles). Nuisance insects and noxious plants would dominate many of our parks and playgrounds in the absence of pesticides. Even fish populations are enhanced by the use of herbicides. By eliminating a portion of weeds that serve as cover for small fish, more are consumed as prey of bigger fish, thus maintaining a healthy fish population. Owners of land near lakes and ponds benefit from pesticide use which allows fuller utilization of the water for safer swimming and boating; their property values are higher if the water is accessible.

Society

Pesticides sustain food production at high levels by protecting crops from pests. The efficiency of the farming community determines the reliability of food production for the nation, and pesticides make it possible for US farmers—who represent less than two percent of the population—to produce enough food for all of us. This leaves the vast majority of citizens free to engage in vocations and professions (the arts, science, education, etc.) that contribute to the public good in other ways.

Duckweed covered this lake before treatment with an aquatic herbicide. These before and after photos (left) show the dramatic difference.

Exports

Pesticides play a vital role in our high agricultural productivity; this leads to a favorable export market, which in turn benefits our balance of trade and offsets the cost of importing foreign oil and other goods.

Transportation

Herbicides are used to eliminate brush and tall weeds around intersections, traffic signs, and guardrails, affording better roadside visibility and reducing accident potential. They also are used to control vegetation along the asphalt shoulder to

Grain is loaded on trucks in the field (upper left) and transported to local elevators (upper right). From there it is delivered to facilities where it is loaded onto barges (above left). Ships carry grain to the world (above right).

Harvey Holt

A traffic sign particularly obstructed by vegetation. Mowing and cutting can be used repeatedly to minimize the problem, but pesticides often are the best answer for long-term control.

Weeds at a railroad crossing (right) can contribute to accidents involving trains and vehicles. Weeds along tracks (above) can cause deterioration of the area around the rail and lead to unsafe track and ballast areas. Excellent weed control along tracks (lower right) provides a clear view of surrounding terrain.

enhance water runoff, thereby reducing the likelihood of accidents caused by hydroplanning and also reducing maintenance costs by extending the life of paved surfaces. Herbicides control weeds and brush along railroad tracks, at crossings, and in railroad yards to increase worker and passenger safety. Wood preservatives protect railroad ties from insect feeding and decay. Airlines apply fungicides to their jet fuel to reduce the growth of mold in the fuel filters.

Utilities

Utility companies use herbicides to suppress the growth of seedling trees that would otherwise interfere with power lines and disrupt electrical service. Stump treatments control sprouting where trees have been cut down. Pipeline crews use herbicides to control vegetation that could envelop aboveground pipes and connections; exposure of the pipes is essential to ward off vandalism and to facilitate maintenance inspections and repairs. Herbicides keep power transmission

A tree stump beneath power lines has been treated with a herbicide to prevent sprouting.

The area around this power station must be kept vegetation free to provide workers clear visibility when working around high voltage lines.

substations from being overgrown by vegetation; they are cheaper and less dangerous than power tree-trimming and assure worker visibility around high voltage power lines.

The Integration of Pesticides with Nonchemical Strategies

The discussion on benefits would be incomplete without mentioning that pesticides are routinely combined with nonchemical pest management strategies. Nonchemical pest solutions are effective in certain situations: planting resistant varieties for apple scab control, mowing or tilling for weed control in problem areas, using pheromones (sex attractants) to lure and capture insects, etc. But, while nonchemical approaches are sometimes cheaper, easier, or more convenient than pesticides, pests often adapt to single system approaches and eventually require a combination of methods to achieve satisfactory pest control. Pesticides are one solution.

State and federal agencies monitor for insects, plants, and plant pathogens that could threaten public health and safety, natural resources, or private property. A current example is the monitoring of mosquitoes to head off an epidemic of West Nile virus. The first outbreak of the virus in the United States occurred in the Northeast in the fall of 1999; since then, it has been carried south and west by birds. West Nile virus has resulted in the deaths of hundreds of people and horses and thousands of birds nationwide.

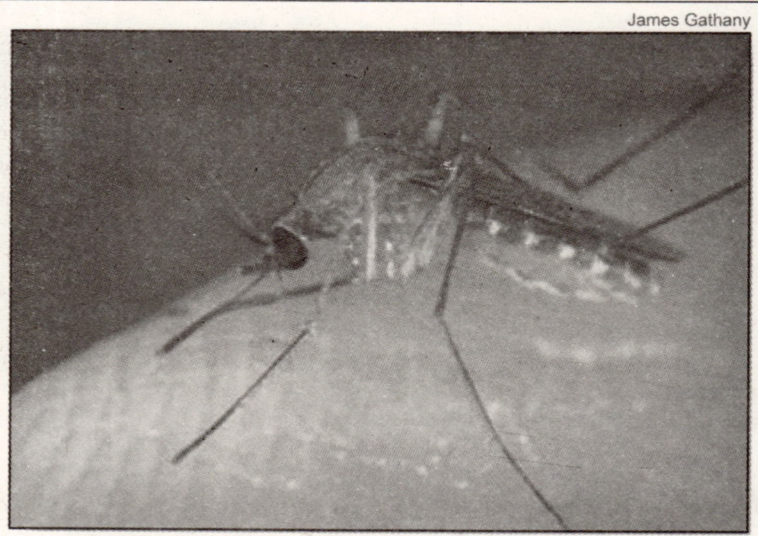

James Gathany

Mosquitoes are vectors for the West Nile virus. Eliminating puddles and small containers of water in residential settings is an effective method used to reduce mosquito populations.

The goal of public health officials is to reduce mosquito numbers to levels that minimize the risk of West Nile virus transmission. Professionals who fight the mosquitoes that carry the virus have long understood that it takes a full arsenal of pest management strategies to gain the upper hand. These include identification and elimination of mosquito breeding sites, use of sentinel species such as crows and blue jays to detect the virus, and the monitoring of adult mosquito populations. Media alerts help citizens protect themselves, and educational programs contribute to control by informing homeowners that they should eliminate prime breeding sites such as standing water in old tires, flower pots, bird baths, and gutters.

This applicator is applying a granular insecticide to stagnant water where mosquito larvae are living.

However, in spite of our best efforts, mosquito populations sometimes explode beyond acceptable public health standards. When this happens, health officials take immediate action to control adult mosquitoes with insecticides in areas where the virus is detected (above). Once the insecticides have controlled the mosquito population, professionals resume their focus on nonpesticidal methods of prevention.

Professionals know that nonchemical systems sometimes fail; and when they do, pesticides are necessary to gain control before the problem becomes intractable.

Whether the pests are cockroaches in a restaurant, flour beetles in kitchen cabinets, or rats in a warehouse, best management strategies include both pesticidal and nonchemical methods for long-term control.

The invasive weed Purple loosestrife grows along a pond (above). Although the flowers are beautiful, the weed tends to take over where cattails normally grow; and since very few wildlife species use purple loosestrife, it is considered detrimental to the "natural" system, especially in wetlands where a diverse native flora is highly desired. Herbicides are effective in controlling purple loosestrife (right).

Conclusion

People who argue against the use of pesticides believe that pest elimination can be achieved without their use. While this may be true in a few isolated situations, most pest management programs around the home, on the farm, aboard planes and trucks, and in parks and natural areas rely on a combination of nonchemical and chemical control methods.

Nonchemical pest management alternatives certainly lessen the need to use pesticides, but they cannot totally eliminate it. Neither pesticides nor nonchemical alternatives such as host plant resistance, timing of planting, painstaking sanitation efforts, etc., offer permanent solutions in all cases. The most effective strategy is an integrated pest management approach.

We must understand both the benefits and the risks of pesticides in developing integrated pest management strategies. Reduced exposure and finely tuned pesticide product selection minimize negative impacts of pesticide use to humans and the environment. Although it may be difficult to convey to the public a sense of balance among pesticide benefits and risks, this issue is central to America's discourse on the continued use of pesticides. Making an informed decision—whether public policy or individual choice—is impossible without evaluating the benefits of use alongside the risks.

Acknowledgments

We thank Stephen Adduci for his original illustrations in The Benefits of Pesticides: A Story Worth Telling. We also thank the following individuals whose constructive comments improved the quality of the publication.

Tom Bauman, Purdue University

Reece Dewell, Purdue University

Bob Ehr, Dow AgroSciences

Fred Fishel, University of Florida

Claire Gesalnan, Environmental Protection Agency

Garry Hamlin, Dow AgroSciences

Steve Hawkins, Purdue University

Robyn Heine, Dow AgroSciences

Dave Johnson, Stewart Agricultural Research Services

Laura Karr, DowAgroSciences

Ken Kukorowski, Aventis

Daniel Kunkel, IR-4

John Obermyer, Purdue University

Paul Pecknold, Purdue University

Mark Peterson, Dow AgroSciences

John Primus, DuPont

Nancy Ragsdale, United States Department of Agriculture

Domingo Riego, Monsanto

Paul Schmitzer, Dow AgroSciences

Marvin Schultz, Dow AgroSciences

Jeff Wolt, Iowa State University

We extend our special thanks to those who have so generously loaned us photographs for this publication. We value their contributions, and we are especially grateful to The William R and Clarice V Spurlock Museum at the University of Illinois for honoring our request to use the ancient Greek letter.

References

Berenbaum M (chairperson). 2000. "The Future Role of Pesticides in US Agriculture". National Research Council.

Council for Agricultural Science and Technology. 2003. "Integrated Pest Management: Current and Future Strategies". Task Force Report No. 140.

Curtis C 1995. "The Public and Pesticides: Exploring the Interface. USDA National Agricultural Pesticide Impact Assessment Program". For full report, go to *http://plantpath.osu.edu/pubpest/*

Environmental Protection Agency, 2000. "The Role of Use-Related Information in Pesticide Risk Assessment and Risk Management". For full report, go to *http://www.epa.gov/oppbead1/use-related.pdf*

Gianessi L and S Sankula 2003. "The Value of Herbicides in US Crop Production. National Center for Food & Agricultural Policy". For full report, go to *http://www.ncfap.org*

Feinberg R, F Whitford and S Rathod 1992. "Perceived Risks and Benefits from Pesticide Use: The Results of a Statewide Survey of Indiana Consumers, Pesticide Professionals, and Extension Agents". For full report, go to *http://www.btny.purdue/PPP*

Johnson W, R Sneda, S Hans and K Hellwig 2002. "Weed Management Systems for Environmentally Sensitive Areas". University of Missouri Cooperative Extension Service Monogram IPM 1018.

Knutson R 1999. "Economic Impacts of Reduced Pesticide Use in the United States: Measurement of Costs and Benefits". Agricultural and Food Policy Center Paper 99-2. For full report, go to *http://www.afpc.tamu.edu*

Padgitt M, D Newton, R Penn and C Sandretto 2000. "Production Practices for Major Crops in US Agriculture", 1990-97. USDA Statistical Bulletin No. 969.

Phillips-McDougall P 2003. "The Cost of New Agrochemical Product Discovery, Development and Registration in 1995 and 2000". For full report, go to *http://www.croplifeamerica.org*

Pike D, F Whitford and S Kamble 1997. "Pesticides and the Bottom Line". Crop Science Special Report Number 1997-10. University of Illinois at Urbana-Champaign.

Smith M, P Blanchard, B Johnson and G Smith 1999. "Atrazine Management and Water Quality: Missouri Guide". University of Missouri Cooperative Extension Service Manual 167.

Van Ravenswaay, E. 1995. "Public Perceptions of Agrichemicals". Council for Agricultural Science and Technology.

Whitford F 2002. "The Complete Book of Pesticide Management: Science, Regulation", Stewardship, and Communication. John Wiley & Sons, Inc.

2

Agricultural Chemicals – How Much Input is Required, How Much is too Much?

Gerd Fleischer

More than 40 years after the onset of the Green Revolution, the use of chemicals in developing country agriculture continues to be intensively debated. How much chemical use is needed? Are there cost-effective alternatives that contribute to rural poverty reduction?

The use of chemicals is one of the most disputed issues in agricultural development. For some, modern technology including a package of synthetic fertilizer, chemical pesticides and high-yielding seed varieties are the key for sustainable development, providing enough food and agricultural raw material for the world's growing population. Increasing yields per unit of land additionally benefit the environment by limiting the encroachment of the agricultural frontier into pristine natural reserves. For critics, chemicals are an unnecessary evil that damages human health and the environment, that accelerates trends to push small-scale farmers out of business, and that hence increases rural poverty. To them, the contribution of chemicals to global production of food and fiber is a myth and the continuous focus on farming methods that depend on high amounts of external inputs is irrelevant for achieving the Millennium Development Goals in general, and the

Source: www.rural-development.de © Entwicklung & Laendlicheraum/Agriculture and Rural Development. Reprinted with permission.

reduction of rural poverty in particular. Resource-poor small-scale farmers in developing countries should rather rely on low external input agriculture to optimize the use of available, but limited resources.

How much chemical use is needed in developing country's agriculture? Far from being a discussion between agricultural experts and advocates of environmental protection only, the debate goes to the heart of the current policy discussions about how to achieve the targets of the Millennium Development declaration. For example, the Copenhagen consensus—a forum of leading development experts—sees agricultural technology, as one of the key components for reducing hunger and malnutrition (*www.copenhagenconsensus.com*). Is there too little or too much emphasis on agricultural technology when we want to reduce hunger and alleviate poverty? And what kind of technology ist needed? Is a world without chemicals a solution?

Chemicals in Agricultural Intensification

The use of synthetic fertilizers and chemical pesticides in developing countries has grown substantially during the past four decades. Governments promoted the use of agrochemicals in order to achieve national food security and increase the production of export crops.

The OECD forecasts that the production and use of chemicals will continue to shift from developed to developing countries. Key developing country regions such as Asia and Latin America show increasingly high growth rates of pesticide use.

Meanwhile, the crop protection industry of the industrialized countries has witnessed a consolidation through acquisitions and mergers to lead to a few global research-based companies. In 1983 there were 27 large and medium-sized research based agrochemical companies, whereas the number has shrunk to eight in 2002. The six largest multinational corporations account for about 85 percent of the total worldwide pesticide sales, currently estimated at US$29 billion per year. Developing countries hold a share of 37 percent (see table).

Global Market for Plant Protection Products in 2000

(In million USD)

Origins	Herbicides	Insecticides	Fungicides	Others	Pesticides, Total	GMOs*	Total
Industrialized Nations	9311	3888	3913	1028	18140	2373	20513
Developing Countries	4483	4121	1888	330	10822	671	11493
Total	13794	8009	5801	1358	28962	3044	32006

* Genetically modified organisms with plant protection properties.

Source: after FAO, 1999.

Increasingly, generic, non-brand, offpatent pesticides become available in developing countries. Those pesticides, especially in the case of insecticides, tend to be broad-spectrum compounds which destruct the balance of the agro-ecosystem by killing beneficial insects that might have otherwise served as natural pest control. Generic pesticides tend to be inexpensive but more toxic for human health and the environment. A considerable share of agricultural pesticides in developing countries is classified as extremely, highly and moderately toxic (WHO Classification Ia, Ib and II). In Ghana, for example, the share of these chemicals in total pesticide use was about 75 percent in the period 1995 to 2000. Given the conditions of use among small scale farmers and workers in developing countries, international donor organizations such as the World Bank have adopted guidelines that effectively ban these pesticides or restrict the supply to trained personnel.

While agricultural intensification is generally considered necessary to meet the projected increase of worldwide demand for food and fiber due to population and income growth, there is a less clear picture about the role of chemicals in these strategies. For example, reviewing available global evidence, researchers of the International Food Policy Research Institute (IFPRI) observed that there is a paradox of increased pesticide use and increased losses from pests (Yudelman *et al.,* 1998). Intuitively, one would expect rather the contrary, i.e., a decrease of losses as a result of increased chemical use. Excessive pesticide use may even result in crop failure at national scale (see Case Study 1 – Pakistan in box).

Case Study 1: Unregulated Pesticide Use Causes Crop Failure in Pakistan

From 1980 to the end of the 1990s, the amount of plant protection products used in Pakistan rose from 665 tonnes to over 44000 tonnes. The markets had been liberalized and product standards relaxed. At the same time, aggressive marketing strategies for insecticides contributed to a sharp increase in the use of plant protection products in cotton production. But plant protection strategies which were not based on sustainable principles caused cotton yields to drop because of attacks by pests and diseases (see graph). Pests quickly built up resistance, which lead many unwitting growers to use more chemicals. Some insects not known as pests in 1980 developed into plagues which were hard to control. Cotton provides directly or indirectly employment for about 40 percent of the total workforce, including farmers, landless female labourers, and employees in textile mills. The negative effects of this «pesticide treadmill» were therefore felt throughout the entire rural economy.

Excessive Fertilizer Use is Widespread

Although, the excessive use of chemical fertilizer in developing countries often receives less attention than the use of pesticides, economic and ecological impacts can be dramatic. Bangladesh faces severe loss of topsoil fertility from overuse of chemical fertilizer and pesticides. Excessive use of nitrogen fertilizer is responsible for contamination of groundwater resources in many countries. Water supply for human consumption has to look for costly relocating of raw water extraction.

Among other emerging economies, China is rapidly moving towards the situation in many developed countries where agriculture has become the main source of water pollution. Seven provinces in the coastal region show a high risk to human and environmental health. It is expected that non-point pollution from crop production will continue to increase and become the major cause of poor water quality, and also an important cause of air pollution. Non-point pollution from crop production is much more difficult to control than point source pollution from livestock. The key issue for the Chinese government is neither the lack of knowledge about nonpoint pollution nor the lack of technologies to control it. A recent task force suggested to set up a policy framework and institutional mechanisms to encourage farmers to adopt available technologies and management practices (*www.harbour.sfu.ca/dlam/newdevelopment.html*).

Apart from the ecological damage, the excessive use of agrochemicals has two main negative economic impacts. First, by causing production costs to be much higher than they need to be, hence by reducing net farm incomes. Secondly, by damaging soil structure through secondary salinization and other forms of physical, chemical and biological degradation that reduce crop yields. Excessive fertilizer use is only one side of the coin. Soil fertility depletion is another major constraint. Research at Wageningen University in 37 African countries has shown that nutrients worth US$11 billion a year at current prices have been lost from cultivated land during the past 30 years. These include a total of 132 million tons of nitrogen, 15 million tons of phosphorus, and 90 million tons of potassium.

Investment in maintaining soil fertility is therefore a priority. An integrated nutrient management strategy is needed that combines soil erosion control, water management, returning crop residues to the soil and additional fertilizer input. A focus on access to chemical fertilizer only will not improve the situation, neither from biophysical nor from economic point of view. The costs of moving chemical fertilizer from a port city to farms in the hinterland are often prohibitively high. It is tempting to call for subsidies to lower the transaction costs for resource-poor farmers who cannot afford to pay for fertilizer at the current rate. But it has been shown in many locations that smallholder farmers in Africa can access nutrients at lower cost, e.g. through agroforestry techniques. Biomass can be transferred from hedges into fields. Leguminous crops in mixed cropping systems make additional nitrogen available.

Pesticide Use and Cotton Yield

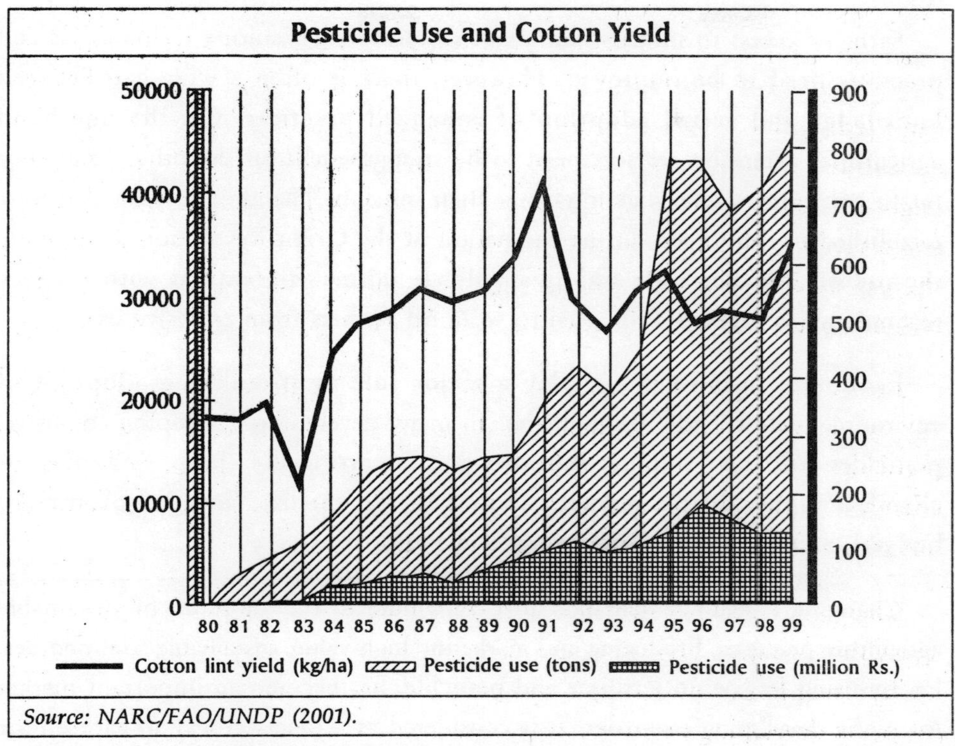

Source: NARC/FAO/UNDP (2001).

Problem Identified – What can be Done?

A range of policy interventions is necessary to create an enabling environment for the adoption of environmentally friendly technologies and practices by farmers. Regulatory policies play an important role in restricting access to the most toxic and damaging chemicals and regulating its use. Developing countries are increasingly pressed to adopt more stringent regulations in order to secure access of their products to global markets (see box: Case Study 2 – Thailand).

Case Study 2: Reform of Pesticide Policy in Thailand

The financial crisis in Asia in 1997 triggered the realization by the Thai government that the high-value export markets for agricultural products, on which the country was heavily dependent, could be at risk. Thai exports had already been rejected several times because they contained pesticide residues. At the same time, domestic consumer demand for food free from such residues rose markedly. Following intensive discussions with all the stakeholders, the government approved a master plan for a revision of the legislation on pesticides, drawn up with support from the GTZ. This master plan subjected the highly liberalized Thai market to stricter controls. Some of the most dangerous chemicals were banned altogether.

Farmers' access to information about sustainable agriculture technologies and practices need to be improved. However, there is often a wide gap between knowledge and actual adoption of enhanced practices. On the one hand agricultural extension services need to be strengthened; on the other hand they might encounter a challenge to change their mission. The same services that were established or reinforced during the period of the Green Revolution to promote the use of agrochemicals among smallscale farmers unfamiliar with modern technology packages, will be used to wean off farmers from excessive use.

Economic incentives can play a major role in stimulating adoption of environmentally friendly technologies. In many developing developing countries, pesticides still carry implicit subsidies and are artificially cheap. Subsidies for chemical inputs distort markets, induce inefficient use, burden government budgets, and increase environmental and health damage.

Changing consumer demands also contribute to the adoption of sustainable agriculture practices. Producing and marketing high-value, sustainable commodities, i.e. by using less or no fertilizer and pesticide, has become an important market for many developing countries. It is considered as a practical way to capture the willingness-to-pay of consumers for environmental and social services of agricultural production through market-based mechanisms.

Food safety and quality standards by importers and retailers have become important driving forces for reducing excessive chemical use in agriculture. The European coffee industry, for example, agreed on a code of conduct for sustainability in coffee production, trade and processing. The code bans the most toxic pesticides from use and aims at progressive implementation of integrated pest management in worldwide coffee production (*www.sustainable-coffee.net*). The organic movement spearheaded the development of non-chemical, but nevertheless highly productive agriculture. Growing consumer demand has contributed to the fact that organic agriculture is a commercial reality, but nevertheless still a niche market. However, a full conversion of developing countries' agriculture to organic practices seems out of reach at present (see box on East Timor).

East Timor: Organic Farming on a National Scale?

In September 1999, following the end of Indonesian rule, agriculture in the former Portuguese colony of East Timor reached a crossroads. The distribution of production inputs such as seeds, fertilizers and plant protection products had broken down, and the markets for agricultural products lay dormant. Farmers were returning to pure subsistence farming. During the subsequent period under UN administration, the idea emerged of focussing the whole country's agriculture permanently on the principles of organic farming. The country was to position itself on the world market under the brand name "East Timor Organic". However, these plans to go without chemical fertilizers and plant protection products were not taken up in this country, which experiences significant food shortages and has only very limited alternatives for generating revenue outside the agricultural sphere. Nevertheless, East Timor did succeed in selling increasing volumes of organic coffee, the country's main foreign currency earner, on markets in North America, Japan and Europe. This was made possible by long-term support from foreign donors, which helped among other things to provide training for producers.

Conclusions

Agrochemicals can contribute to increased agricultural productivity if their use is limited and targeted. But considerable damage may be the result when users have limited knowledge or where there are incentives for excessive use. When using chemicals, positive impacts should clearly outweigh the negative ones. This calls for the optimal level of chemical use which is subject to local circumstances. In any case appropriate framework conditions are needed to support it.

Unfortunately, many farmers in developing countries, especially the emerging economies tend to emulate the practices of their colleagues in the developed world who continue to strive for short-term maximum yields with the help of excessive chemical use instead of focusing on more sustainable practices.

According to the UN Food and Agricultural Organization, however, it seems possible that the overuse of pesticides and fertilizers will diminish, as will their environmental impacts, because of better technologies, regulatory measures and the growing emphasis on organic agriculture (FAO 2000). To fulfil this promising scenario, a set of policies and institutional changes to facilitate dissemination and uptake of environmentally friendly technologies and practices by the rural population is required. In recent years, consumer market changes have proven to be effective triggers for policy reform.

(Dr. Gerd Fleischer, Deutsche Gesellschaft für Technische, Zusammenarbeit (GTZ) GmbH, Agriculture, Fisheries and Food, Eschborn, Germany, can be reached at Gerd.Fleischer@gtz.de).

3

ICRA Sector Analysis: Insecticides/Pesticides

ICRA sector analysis (insecticides/pesticides) report opens our eyes to the issue of acute food contamination through the use of pesticides, especially in developing countries. According to estimates, such contamination arises 13 times more often in developing countries than in developed countries. The reasons for higher pesticide residue contamination in developing countries are because pesticides are used more indiscriminately here and the security instructions and necessary protection measures not respected. Further, insufficient training and information on the risk of pesticide use and lack of MRLs and other yardsticks measuring risk of pesticide use add to the flippancy in pesticide use. According to a study in India, residue of DDT and benzene hexachloride, both suspected carcinogens, were found in breast milk samples collected from mothers in Punjab.

Structure of the Industry

Overview

A pesticide is any substance or mixture of substances intended for preventing, destroying, repelling, or mitigating any pest. The pesticides (of which insecticides

Source: www.icraindia.com, ICRA Information, Grading and Research Service.

constitute an important segment) or the agrochemicals industry (hereinafter referred to as the PAC industry) primarily consists of insecticides, herbicides and fungicides.

While farmers have been doing pest management since the onset of agricultural production, the discovery of the pesticidal properties of synthetic chemicals in the middle of the 20th century has transformed agriculture. DDT was discovered as an insecticide in 1939 and was used extensively in agriculture, and for public health programs. Subsequently, other insecticides, herbicides, and fungicides were developed. Their large-scale use started in farming in industrialized countries, resulting in large increases in production and/or cost savings on labour. The first Green Revolution, which began in the 1960s, made high-yielding crop varieties available to developing countries, especially in Asia. Exploitation of the production potential of these varieties stimulated the use of fertilizer and pesticides. This dynamic led to large increases in food production in many countries but also to growing pesticide use.

PACs provide vital inputs to crop protection and act as a vital input to agricultural produce. In a world-wide study of eight crops, it has been estimated that the global harvest losses due to pests was about 42% of attainable production. Higher yields can be assured by reducing crop losses caused by weeds, pests and insects. PAC have assisted in controlling pests and maintaining the availability of low cost and high quality food. They also allow for improved storage and distribution of crops, fruits, and grains.

The importance of PACs in India derives from the fact that India is an agrarian society. Agriculture is the backbone of the Indian economy and contributes 22% of India's GDP. Nearly 70% of the country's workforce derives its livelihood from agriculture. Higher crop productivity can be achieved through high-grade crop protection, and the challenge is to prevent or reduce pest related crop losses, which are presently estimated at around 20-25% of crops sown, i.e., approximately Rs.250-300 billion per annum.

Pesticide use significantly contributes for enhancing agricultural production and also helped to reduce the problems of vector-borne diseases. On the other hand, failure to adhere to the safety norms at various stages of pesticide production and use and also some times non-availability of credible information have caused serious concerns in the society. There are health risks associated with pesticides. By their very nature, most chemical-based PACs create some risk of harm to humans, animals, or the environment because they are designed to kill or otherwise adversely affect living organisms. These risks include on-farm ingestion by workers, discharge of toxic chemicals into the air and water, and consumption of foods that contain pesticide residues by consumers. Laboratory studies show that pesticides can cause health problems, such as birth defects, nerve damage, cancer, and other effects that might occur over a long period of time. However, these effects depend on how toxic the pesticide is and how much of it is consumed. Some pesticides also pose unique health risks to children. At the same time, pesticides are one of the vital ingredients of crop protection, and enhancing agricultural productivity.

Categories of Pesticides

Pesticides are often referred to according to the type of pest they control. In such a classification, pesticides are normally categorised into the following major categories:

- Insecticides act against insects which feed on crops, leaves, roots, and other parts of plants;
- Herbicides (also known as weedicides) act against weeds or unwanted plants compete with the crop for nutrients, light, water, space;
- Fungicides act against bacteria, fungi, virus and mycoplasma which cause various diseases in plants.

Other categories of pesticides include:

- Algicides control algae in lakes, canals, swimming pools, water tanks, and other sites.
- Antifouling agents: kill or repel organisms that attach to underwater surfaces.
- Antimicrobials kill microorganisms (such as bacteria and viruses).
- Fumigants produce gas or vapor intended to destroy pests in buildings or soil.
- Miticides kill mites that feed on plants and animals.
- Ovicides kill eggs of insects and mites.

- Repellents repel pests, including insects (such as mosquitoes) and birds.
- Rodenticides control mice and other rodents.

Another way to classify PAC is to consider those that are chemical pesticides or are derived from a common source or production method. Other categories include biopesticides, antimicrobials, and pest control devices.

Chemical Pesticides comprise the following:

- Organophosphate Pesticides were developed during the early 19th century, but their effects on insects, which are similar to their effects on humans, were discovered in 1932. Some are very poisonous (they were used in World War II as nerve agents). However, they usually are not persistent in the environment.
- Carbamate Pesticides.
- Organochlorine Insecticides were commonly used in the past, but many have been removed from the market due to their health and environmental effects and their persistence (e.g. DDT and chlordane).
- Pyrethroid Pesticides were developed as a synthetic version of the naturally occurring pesticide pyrethrin, which is found in chrysanthemums. They have been modified to increase their stability in the environment. Some synthetic pyrethroids are toxic to the nervous system.

Biopesticides are pesticides derived from such natural materials as animals, plants, bacteria, and certain minerals. The advantages of biopesticides over chemical pesticides are: (a) they are inherently less harmful; (b) they are more target specific than chemical pesticides affecting only the target pests and their close relatives. In contrast, chemical pesticides often destroy friendly insects, birds and mammals; (c) they are often effective in small quantities; and (d) they decompose quickly and do not leave problematic residues. Biopesticides can be classified into three major classes:

- Microbial pesticides consist of a microorganism (e.g., a bacterium, fungus, virus or protozoan) as the active ingredient. Microbial pesticides can control many different kinds of pests, although each separate active ingredient is relatively specific for its target pests. For example, there are fungi that control certain weeds, and other fungi that kill specific insects. The most widely used microbial pesticides are subspecies and strains of Bacillus thuringiensis, or Bt. Each strain of this bacterium produces a different mix of proteins, and specifically kills one or a few related species of insect larvae. While some Bt.'s control moth larvae found on plants, other Bt.'s are specific for larvae of flies and mosquitoes. The target insect species are determined by whether the particular Bt. produces a protein that can bind to a larval gut receptor, thereby causing the insect larvae to starve. Other commonly used biopesticides include Baculoviruses and neem. Baculoviruses are target specific viruses, which can infect and destroy a number of important plant pests. They are particularly effective against the lepidopterous pests of cotton, rice and vegetables. Their large-scale production poses certain difficulties, so their use has been limited to small areas. They are not available commercially in India, but are being produced on a small scale by various Integrated Pest Management (IPM) centres and state agricultural departments. Neem contains several chemicals, including 'azadirachtin', which affects the reproductive and digestive process of a number of important pests. Recent research carried out in India and abroad has led to the development of effective formulations of neem, which are being commercially produced. As neem is non-toxic to birds and mammals and is non-carcinogenic, its demand is likely to increase. However, the present demand is very small.

- Plant-Incorporated-Protectants (PIPs) are pesticidal substances that plants produce from genetic material that has been added to the plant. For example, scientists can take the gene for the Bt. pesticidal protein, and introduce the gene into the plant's own genetic material. Then the plant, instead of the Bt. bacterium, manufactures the substance that destroys the pest.
- Biochemical pesticides are naturally occurring substances that control pests by non-toxic mechanisms. Conventional pesticides, by contrast, are generally synthetic materials that directly kill or inactivate the pest. Biochemical pesticides include substances, such as insect sex pheromones, that interfere with mating, as well as various scented plant extracts that attract insect pests to traps.

The Indian PAC industry primarily comprises of insecticides, fungicides, and herbicides/weedicides. Insecticides dominate consumption with around 61% of estimated consumption of PAC, followed by fungicides (18%), herbicides/ weedicides (16%), and others (5%). Insecticides are used mainly for rice, cotton and vegetables; herbicides for rubber, oil palm, tea, and coffee; and fungicides for tobacco, vegetables and bananas.

The per hectare consumption of pesticides in India is estimated at 0.5 kgs. However, this is not uniform. It varies vastly across the country with the agro-ecological settings, cropping pattern, irrigation facilities, intensity of pests and diseases, resistance and resurgence of insect pests, etc. Pesticide use is particularly high in regions with irrigation facilities and also in those areas where commercial crops are grown.

Cotton, paddy, vegetables and fruits are grown in 32% of the cultivated area and account for over 80% of the pesticide consumption in the country. While cotton is planted on about 4.5-5% of the total cultivable area (on about 7.5 million hectares or million ha), it accounts for about 45% of pesticide consumption in India, followed by rice (23%), jowar (9%), vegetables (7%), wheat (6%), and pulses (4%).

At present, 181 pesticides have been registered in India. A Registration Committee (RC) has been constituted under Section 5 of the Insecticides Act, 1968 to register insecticides after scrutinising formulae, verifying claims of efficacy and safety to human beings and animals, specify the precautions against poisoning and any other function incidental to these matters. To assess efficacy of the insecticides and their safety to human beings and animals, the RC has evolved exhaustive guidelines/data requirements which *inter-alia* includes residue in crops on which the insecticides are intended to be used. The onus lies with the importers/ manufacturers to generate data relating to the insecticides for which registration are sought. The Government of India (GoI) has been reviewing the continued use of the registered pesticides at periodical intervals by constituting expert committees. Based on the recommendation of these expert committees, the GoI has refused registration for 18, banned 27 technical grade pesticides, 4 formulations and restricted the use of 10 pesticides in the country. However, a large number of pesticides that fall under the extremely hazardous category of 1A & 1B of World Health Organisation (WHO) classification are yet to be phased out or banned despite recent reports linking some of them responsible for serious health hazards like cancer, congenital malformation, abnormality in reproductive system etc. Such pesticides, still in use include (a) Extremely Hazardous—Aldicarb, Methomyl, Monocrotophos, Phorate, Phosphomidon; and (b) Highly Hazardous—Carbofuran, Dichlovos, Dimethoate, Endosulphan, Ethion, Fenthion, Fenvalerate, Methyl Parathion, Phosalone, and Triazophos.

Global Market for Pesticides

The global market for chemical pesticides/agrochemicals was estimated at around US$26.71 billion in 2003. Herbicides comprised 44% of the world market, followed by insecticides (27%), fungicides (20%), and others (9%). The market increased by 6.2% during 2003. In the period since 1990, the world market has increased from US$23.17 billion, at an annual average of 1.1%.

Market conditions that prevailed in 2003 were far more favourable than in the recent past, and if it had not been for the drought that affected Northern European countries, the industry would have recorded higher growth. Apart from the drought in Northern Europe, the major factors that affected the market were a general improvement in commodity prices (except in rice) that benefited farm incomes in the major developed markets; significant economic improvement in Latin America that gave farmers increased access to credit and hence ability to purchase agrochemicals; recovery from drought in Canada, India, Australia and much of Asia, and the 2002 Farm Act in the US and the mid-term common agricultural policy (CAP) reform in the European Union (EU) that gives farmers a greater surety in their income position in the future. On the negative side, increasing generic competition and price reductions reduced values; further rice acreage reduction and a competitive market situation resulted in a continuation of the market downturn in Japan and a continued increase in GM acreages resulted in value being taken from the conventional crop protection market, particularly the herbicides sector.

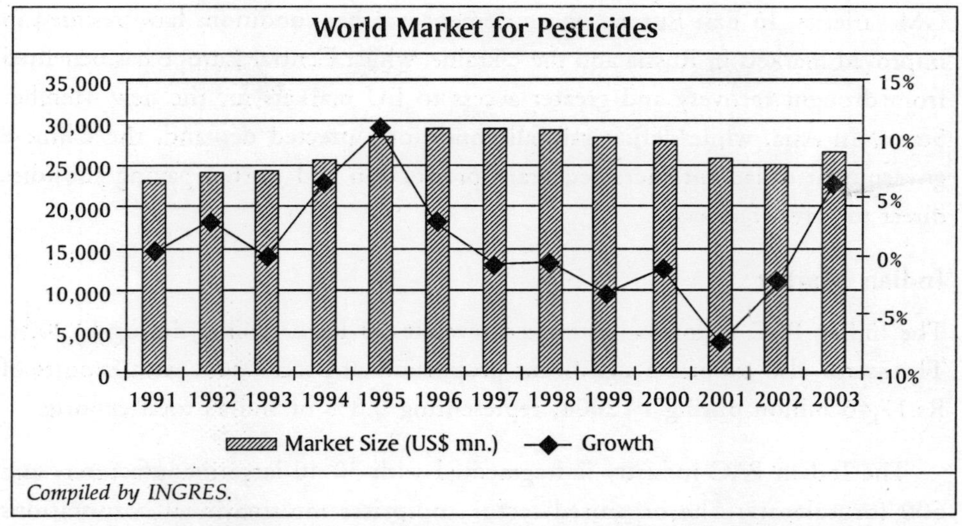

Compiled by INGRES.

The markets for chemical pesticides have undergone rapid changes over the last decade. Overall pesticide use worldwide over the last decade has remained constant or declined. The reduction can be explained partly by changing crop prices, greater efficiency of pesticide use as a result of improvements in pest management practices and technology, increased adoption of genetically modified (GM) crops, and IPM techniques aimed at both improving pest management practices, and in some cases targeting a reduction in pesticide use.

In 2003, the market for GM crops increased by 19% to reach US$3.94 billion. The majority of these sales continue to be derived from North and Latin America, although uptake in other regions is increasing, particularly in China and India. The focus of GM crop development is in the soybean, maize, cotton and canola sectors, however in many of the countries where GM varieties have been accepted, the area cultivated with these varieties is now approaching maturity.

The global agrochemicals market is estimated to have recorded positive growth during 2004, due to recovery from some of the negative factors that affected the market in 2003. In Northern Europe, a return to more normal weather conditions has benefited agrochemical usage. In Latin America, economic recovery continues to re-invigorate the agricultural economy, with agrochemical sales also benefiting from the soybean rust outbreak. In North America, planted acreages of maize, soybean and cotton have all risen, although much of the increase has been with

GM varieties. In East Europe, more clement weather conditions have resulted in improved markets in Russia and the Ukraine, whilst Central Europe has benefited from drought recovery and greater access to EU markets for the new Member States. In Asia, while deficit rainfall conditions affected demand, the Chinese government called for increased grain production and started paying subsidies direct to cereal farmers.

Indian Market

The Indian PAC sector had estimated revenues of Rs.40 billion during FY2004. The sector also exports a significant proportion of production, with exports of Rs.17.46 billion during FY2004, representing 0.8% of India's total exports.

The Indian PAC industry is fragmented with 30-40 large manufacturers and 600 formulators. The organised sector comprises multinational corporations (MNCs) and Indian players. Besides the presence of MNCs such as Monsanto, Bayer, and Syngenta, there are many major Indian players such as United Phosphorous, Excel Industries, and Rallis India.

Traditionally, the Indian players in the PAC market have focussed on marketing generic and off-patent products, whereas the MNCs have focused on high-end specialty products. The Indian companies are playing to their basic strengths, which are low-cost manufacture, applied research and extensive distribution. Indian players focus on applied research, which include developing processes to manufacture off-patent products, more effective methods of delivering existing products, and new formulations (or combinations) of generic products. They generally do not have the financial and technical resources for undertaking basic research. The MNCs dominate the market for proprietary products (patented new molecules). They have the advantage of access to the R&D pipeline and superior financial resources of their parents. Despite the threat of generic products, MNCs have introduced quite a few of their top selling global products in the Indian market. Monsanto holds Indian registrations for almost the entire complement of global brands. The Indian operations of MNCs have also tried to guard against the cyclicality of the domestic market, by identifying and sticking to niche categories, crops and markets within the industry. MNCs have also been working at strengthening the linkages with the farmers through extension and advisory services.

Many MNCs have also entered into supply arrangements with their parents to take advantage of the cost advantage in manufacturing from the Indian facility. Due to medium capital intensity (Operating Income/assets of 1), and Government policy of encouraging the small-scale sector, intense competition exists. Distribution reach, new product launches, and ability to extend credit to the farmers influence sales.

The onset of the product patent regime in 2005 can lead to relatively better prospects for the MNCs in the business. In the post-2005 scenario, the Indian market can expect to witness an inflow of new molecules from MNCs. The domestic industry would have to meet this competition through development of new molecules, process and product development, development of new formulations and by better product customisation. Almost 70% of all agrochemicals present in use in India are off patent. In the new product patent regime, in which MNCs are expected to bring in new molecules, the domestic players would need to focus on technology development for bio- and botanical-pesticides.

Key Issues Facing the Players

Cyclical Nature of Demand

Overall pesticide usage by farmers depends on a multitude of factors, such as climatic conditions, the area under cultivation, composition and variety of crops, pest and disease pressures, farm incomes, pesticide cost/crop price ratios, pesticide policies and management practices.

The demand for PAC in India is seasonal and cyclical as it largely depends on agricultural production. PAC demand is skewed in favour of kharif crops, with kharif crops accounting for around 70% of annual consumption of PAC. The peak consumption of PAC is during July-November. As the business is highly seasonal, PAC companies have to build up inventory in advance of the season by making assumptions about the monsoon and crop acreage situation.

As stated above, overall PAC consumption primarily depends on agricultural production, which in turn is highly dependent on rainfall. The ratio of net irrigated area to net sown area is around 39% in India. Droughts/deficient monsoons not only adversely impact kharif sowing season (beginning in July) but also kharif production. They also impact water reservoirs and rabi production in the winter. Adverse monsoons also impact the loan repaying ability of farmers. Apart from agricultural production growth, consumption of PAC also depends on the pest pressure. If weather conditions are such that there are more insects, then farmers are likely to use more insecticides. Even if monsoons are adequate, the attack of insects or diseases may or may not be very high. Demand growth is a very complex result of various weather conditions, and unpredictable events like pest pressure and attack of disease also impact PAC consumption.

PAC consumption is estimated to have declined during FY2003. During FY2003, Indian agriculture experienced one of the worst droughts in 2002, with the South-West monsoon season being at only 81% of the Long Period Average (LPA). The sowing month of July, which normally receives 30% of the season's rain, suffered a shortage of about 49%. Area under kharif-production of foodgrains declined from 73.9 million ha in FY2002 to 67.3 million ha in FY2003. Area under rabi-production of foodgrains also declined from 48.1 million ha to 45.8 million ha. Area under cotton production also declined from 9.1 million ha in FY2002 to 7.5 million ha in FY2003. As a result, PAC production declined 16.7% during FY2003 to 67,400 MT, because of drought resulting in lower kharif and rabi output during FY2003. Production of foodgrains and non-foodgrains was adversely affected, with the fall in kharif output sharper than that of rabi. While kharif production of foodgrains declined from 112.1 million tonnes in FY2002 to 87.8 million tonnes in FY2003, rabi foodgrain production declined from 100.8 million tonnes to 86.4 million tonnes.

Area Under Cultivation in India of Major Pesticide-using Agricultural Products

	Area under cultivation – million ha							CAGR (%)	
FY	1998	1999	2000	2001	2002	2003	2004	1991-2000	2001 to 2004
Cotton	8.87	9.34	8.71	8.53	9.13	7.67	7.64	2.36%	-3.61%
Rice	43.45	44.80	45.16	44.71	44.90	40.28	42.41	0.62%	-1.75%
Wheat	26.70	27.52	27.49	25.73	26.34	24.86	26.62	1.67%	1.14%
Jowar	10.80	9.79	10.25	9.86	9.80	9.20	9.49	-3.71%	-1.27%
Arhar	3.36	3.44	3.43	3.63	3.33	3.38	3.53	-0.65%	-0.93%

Compiled by INGRES.

However, PAC demand increased to around 49,000 tonnes during FY2004, because of excess rainfall during kharif sowing season beginning July 2003. Foodgrains production increased 22.6% during FY2004 to 213.5 million tonnes. While kharif production of foodgrains increased from 87.8 million tonnes in FY2003 to 116.9 million tonnes in FY2004, rabi foodgrain production increased from 86.4 million tonnes to 96.6 million tonnes. The encouraging factor was the high produce price for cotton, paddy and several other crops that the farmer was able to realise. However, since the inventory of pesticides in several key markets is higher compared with normal levels, performance in the kharif season was muted.

During FY2005, PAC consumption and demand is expected to have declined, because of deficient rains. The cumulative area-weighted rainfall during the South-West monsoon was 13% below the LPA. However, as the total area of the sub-divisions experiencing drought conditions was only 18%, the year was not declared as an all-India drought year. According to the Third Advance Estimates released by the Ministry of Agriculture, kharif foodgrains production in FY2005 is expected to decline to 104.1 million tonnes, mainly because of lower production of coarse cereals. The satisfactory beginning of the North-East monsoon coupled with a comfortable water reservoir position augured well for rabi sowing. Rabi foodgrains production is expected to increase to 106.4 million tonnes during FY2005, thereby recouping some of the loss of kharif output.

Pesticides Consumption and Agricultural Production

FY	1998	1999	2000	2001	2002	2003	2004	2005
PAC consumption-MT	52,239	49,157	46,195	43,584	47,020	48,000	49,000	
Agricultural Production Growth	-5.8%	7.6%	-1.3%	-6.3%	7.6%	-15.2%	19.6%	
Foodgrains-million tonnes	192.3	203.6	209.8	196.8	212.9	174.2	213.5	210.4
Of which rice	82.5	72.7	77.5	72.8	93.3	72.7	88.3	87.1
Wheat	66.4	71.3	76.4	69.7	72.8	65.1	72.1	74.1
Non-foodgrains								
Of which cotton	10.9	12.3	11.5	9.5	10.0	8.7	13.9	16.1

Compiled by INGRES.

As shall be discussed below, the overall pesticide use in India over the last decade has remained constant or declined. This can be explained partly by changing crop prices, greater efficiency of pesticide use as a result of improvements in pest management practices and technology, and government policies aimed at both improving pest management practices, and in some cases targeting a reduction in pesticide use.

Because of declining domestic consumption, production has also declined in recent years. As compared with capacity of 145,000 MT during FY2004, overall capacity utilisation was 47% during FY2004 (44% in insecticides, 37% in herbicides/weedicides, 76% in fungicides, and 43% in other pesticides).

Increased Reliance on Exports

Because of the cyclical nature of PAC consumption in India, and the long-term trend of declining demand, the PAC industry in India has resorted to increased exports for better capacity utilisation.

Pesticide Production, Consumption, Exports and Imports

Thousands of tonnes

	FY2000	FY2001	FY2002	FY2003	FY2004
Production	95.33	92.27	81.80	67.97	68.00
Domestic Consumption	46.20	43.58	47.02	48.00	49.00
Exports	40.65	47.92	50.09	55.69	68.98

The export market often provides a cushion to the vagaries of the domestic market. Out of the combined turnover of Rs.40 billion, the export turnover is around Rs.17.5 billion. Companies such as United Phosphorous, Excel Industries, etc., are significant players for generic crop protection products. An export-oriented strategy for the Indian PAC has the potential to deliver rapid expansion in volumes and provide better insulation from the cyclicalities of the domestic market. While exports increased 23.9% during FY2004 in volume terms to 68,981 MT, the growth in value terms was 17.3%. However, because of increased domestic demand, imports increased from 6,795 MT during FY2003 to 12,647 MT during FY2004. During FY2004, insecticides accounted for 71% of exports in volume, and 83% of exports in value.

Exports of Pesticides												
	Exports-MT				Exports-Rs. million				Realisations-Rs./kg			
FY	2002	2003	2004	2005 (9M)	2002	2003	2004	2005 (9M)	2002	2003	2004	2005 (9M)
Insecticides	37,816	42,978	49,072	35,255	12,197	13,305	14,562	10,751	323	310	297	305
Fungicides	7,515	7,682	12,083	10,568	722	814	1,362	1,558	96	106	113	147
Herbicides	2,237	2,093	3,032	2,641	381	436	424	383	170	208	140	145
Others	2,517	2,940	4,794	3,601	264	320	1,107	718	105	109	231	199
Total	50,085	55,693	68,981	52,066	13,565	14,875	17,456	13,410	271	267	253	258

Compiled by INGRES.

The Indian PAC industry exports primarily to US, Argentina, France, Brazil, and Netherlands. The US was the largest market for Indian pesticide exports during FY2004, accounting for 9.1% of total exports, followed by France (7.9%), and Netherlands (7.5%).

Export Growth of PAC

(Rs. million)

	FY2000	FY2001	FY2002	FY2003	FY2004	FY2005 (9M)	3-year CAGR
France	654	1,100	1,161	1,451	1,377	1,266	7.8%
US	810	1,027	1,140	1,684	1,592	970	15.7%
Netherlands	1,569	1,636	1,543	944	1,309	807	-7.2%
Brazil	146	392	291	606	969	741	35.2%
Belgium	395	275	386	540	700	705	36.5%
Argentina	200	260	406	1,018	521	568	26.1%
Others	6,277	7,465	8,638	8,632	10,988	8,352	12.4%
Total	10,050	12,153	13,565	14,875	17,456	13,410	12.8%

Compiled by INGRES.

The increased exports of pesticides in recent years is primarily because of the reduction in pesticide production in developed countries, and the shift in pesticide production from the developed countries to developing countries.

India is one of the largest producers of pesticides in Asia. Total world pesticide exports from all countries increased from US$10.27 billion in 2002 to US$12.42 billion in 2003, caused by higher pesticide usage. Amongst the developing countries, India is the second-largest exporter of pesticides, behind China.

India accounted for 3% of the world export of pesticides in 2003, as compared with 5.9% for China. In terms of market share of exports of various pesticide products, Indian exports of insecticides aggregated US$313 million during 2003, accounting for 10.4% of total worldwide insecticide exports of US$3,011 million. India's share has declined from 11.4% during 2002.

Major World Exporters of Pesticides

	Value of Exports – US$ Million			Share of World Exports		
	2001	2002	2003	2001	2002	2003
France	1,477	1,573	1,861	13.9%	15.3%	15.0%
Germany	1,510	1,539	1,796	14.2%	15.0%	14.5%
US	1,547	1,547	1,457	14.5%	15.1%	11.7%
UK	981	1,023	1,098	9.2%	10.0%	8.8%
China	549	592	730	5.2%	5.8%	5.9%
Switzerland	598	460	696	5.6%	4.5%	5.6%
India	288	306	375	2.7%	3.0%	3.0%
Others	3,695	3,231	4,410	34.7%	31.5%	35.5%
Total	10,644	10,271	12,423	100.0%	100.0%	100.0%

Compiled by INGRES from UN Data.

Over the years, Indian PAC companies have developed process technologies for many PACs. The easy availability of raw materials, low-cost trained and skilled workforce, low overheads, and technically qualified managerial base have made India an attractive sourcing destination for global MNCs. Many overseas companies are also undertaking collaborative research with local companies and institutions. For instance, United Phosphorous has manufacturing arrangements with Syngenta for supply of some of the generic products. Some Indian operations of MNCs have also entered into supply arrangements with their parents to take advantage of the cost advantage in manufacturing from the Indian facility. India has emerged as a low-cost sourcing base for generic PAC products. Companies have sought to exploit this advantage by exporting generic products to destinations such as Latin America, Africa and, more recently, to the US and Europe. These companies typically look to becoming early entrants into a proprietary product (one patented by an MNC) that is going off-patent in a target market, and try to offer a lower-priced alternative to an existing proprietary product.

Recently, the International Finance Corporation (IFC) approved a US$17.5 million loan to 'United Phosphorous' towards expansion of production capacity, strengthening of financial position by extending its debt maturities, and acquiring product registration rights in the developed markets. As the cost of R&D has increased, demand has stagnated, and patents have expired on many active ingredients, there has been a rapid growth in business of companies producing generics. To users, the advantage of generics is their low price and sales have grown to account for 12-15% of the world pesticide market. According to an Agro report on 64 generic manufacturers, a major proportion probably began production in the last 5-10 years.

Around 65% of the global market for PAC products (by value) is slated to be off-patent by 2005. For Indian companies, which have a negligible share in this market, this market offers large possibilities for increasing exports.

A major threat for PAC exports is regulatory barriers. Access to foreign markets is restricted through registration procedures stipulated by different countries. A company desirous of marketing a generic product in a new market is required to obtain registration for that product in each target market. Registration involves significant initial investments (US$3-10 million per registration in most markets), and long gestation periods of 4-5 years (for field trials, if required). Therefore, companies that have registrations are protected by strong entry barriers, once they establish a presence in the target market. In nearly all countries, pesticides registration schemes exist and approval before marketing is required. The data package required for pesticides is much larger than for new industrial chemicals because pesticides are known to be biologically active and to result in direct exposure of the environment and possibly humans. Over the past several years, the environmental risks of pesticides have been considered much more extensively in granting marketing approval, and they have played a greater role in the decision-making process. Also, more attention is being paid to ensure that there is a real 'added-value' in terms of the efficacy of new pesticides. The development and registration of safer pesticide products is encouraged in many countries through a variety of mechanisms, including reduced data requirements for low-risk biological pesticides, reduced registration fees and commitment to faster registration. Registration can also be used as a non-tariff barrier. Thus, Indian exports to the

European Union (EU) have stagnated over the last few years, because the EU decided to re-register all products. From July 2003, the re-registration process is expected to increasingly affect the market for older products. Whilst there are obvious alternatives for a number of these molecules, opportunities exist for replacement of others in the marketplace. Many countries supplement their pesticide registration programmes with re-registration programmes to bring the test data on older pesticides up to modern standards. Re-registration has high priority in many countries because knowledge about possible adverse effects has grown considerably during the last three decades, and data requirements and hazard evaluations have become much more comprehensive. Information made available through re-registration programmes has been used by some countries to ban pesticides that were once widely used, such as DDT, aldrin, and dieldrin.

Shift in Government Policies in Recent Years

The PAC industry is highly sensitive to Government's policies. The industry is regulated by two ministries. While the Department of Chemicals and Petrochemicals under Ministry of Chemicals and Fertilisers promotes production of pesticides; the Ministry of Agriculture monitors the quality and supply of pesticides. Since pesticides are toxic and hazardous to humans and environment, and also enter into the food chain, the (GoI) regulates manufacture, sale, transport, export/import under Insecticides Act, 1968. Under the Act, no pesticides are allowed for production/import without registration.

The PAC industry developed in India because of the GoI's policies of self-reliance across the board, protection of domestic industry, and assured demand because of the large agricultural production. The GoI has also been framing various policies towards financing of farmers. Effective 1998-99, banks have been issuing Kisan Credit Cards to farmers on the basis of their land holdings so that the farmers can use them to readily purchase agricultural inputs such as seeds, fertilisers, and pesticides and draw for their production needs. Over the last couple of years, the Kisan Credit Card Scheme has emerged as an effective tool for catering to the short-term credit requirements of the farmers.

In recent years, because of increased self-reliance in PAC production, the GoI has reduced customs duties on imports of PAC. While the Union Budget 2003-04 resulted in reduction in customs duty to 25% on all pesticides, the Mini-Budget reduced peak customs duty to 20%. The Union Budget for 2005-06 reduced the peak rate of customs duty to 15%. In the Union Budget for 2000-01, the GoI also increased excise duty on pesticides from 8% to 16%.

In recent years, the PAC industry has diversified into manufacture of biopesticides and synthetic pesticides. Biopesticides are becoming increasingly popular and often are safer than traditional chemical pesticides. In India, after the GoI decided to ban hexachlorobenzene (HCB) in April 1997, an estimated 300,000 tonnes of HCB has been eliminated from use. HCB represented 30% of India's total PAC consumption. The void created by HCB did not impact consumption or export, as biopesticides, botanical pesticides, and synthetic pyrethroids, replaced HCB.

Concern about the adverse effects of chemical pesticides due to their indiscriminate use is growing. Pesticides residues are being found increasingly in Indian farm produce posing a threat to human health. Because of the increased realisation of the harmful impact of pesticides, the GoI promoted the Integrated Pest Management (IPM) in 1985, as an eco-friendly strategy of pest containment by exploiting the role of natural agents/forces in harmony with other pest management practices. The GoI has introduced Integrated Pest Management (IPM) at the central and state levels for purposes of plant protection. IPM was adopted as a national policy in 1995 and it was implemented as large-scale demonstrations cum training in farmer fields in a wide range of crops including cotton. As a major policy change, the GoI also withdrew the subsidy component on pesticides since 1994-95 to facilitate promotion of IPM in the country. In India, 26 central IPM centres have been established in many states for pest surveillance and monitoring, promotion of bio-control methods of conservation, promotion of non-chemical methods of pest control, training of extension workers and farmers, etc. *IPM is discussed in further detail below.* The 10th Five-Year Plan envisages a strengthening of IPM infrastructure, especially for surveillance and forecasting the outbreak of pests and diseases and production/multiplication of bio-control agents for field use. Besides, reliable methods of forecasting are expected to be developed and efforts would be made to make bio-control agents available on demand to farmers to help them adopt IPM. IPM has contributed significantly to the decline in consumption of chemical pesticides.

The consumption of biopesticides has also increased from 219 MT during FY1997 to 902 MT during FY2002. Various steps taken by the GoI to promote the usage of biopesticides include simplification of the guidelines for registration of biopesticides; encouraging farmers, local entrepreneurs, NGOs for production of biopesticides; provision of assistance as grants-in-aid for research, development, and production; provision of grants-in-aid provided to the States for infrastructural development for production of biocontrol agents and biopesticides; and allowing commercialisation of biopesticides during the validity of provisional registration for 2 years.

The GoI is also encouraging lower PAC consumption because of safety regulations on import of food products in various countries (which regulate the quantity of pesticide residues). High level of pesticide residues can often act as a constraint against increased agricultural exports. Rejection of even one shipment because of the discovery of an unknown pest at a port or high pesticide residues can result in the exporting country being placed on a quarantine list for that commodity, thus eliminating one import market. Repeated violations of residue requirements can result in automatic detention (inspection or fumigation or both) of all shipments from a country until it can document sufficient pre-inspection quality control.

The GoI has also recently encouraged trial and production of genetically modified (GM)/transgenic crops *(discussed in detail below)*. Worldwide adoption rates for transgenic crops during the period 1996 to 2004 are unprecedented and are the highest for any new technologies by agricultural industry standards. High adoption rates reflect farmer satisfaction with the products that offer substantial benefits ranging from more convenient and flexible crop management, higher productivity and/or net returns per hectare, social benefits, and a cleaner environment through decreased use of conventional pesticides, which collectively contribute to a more sustainable agriculture.

India has recently approved the planting of a GM-variety of insect-resistant biotech cotton (Bt. Cotton) that reduces the need for pesticides. India commercialised Bt. cotton for the first time in 2002. Only a mere 80,000 acres was cultivated with Bt. cotton in 2002. However, while area increased by 100% in 2003, and 400% in 2004; GM crop area (as percent of acreage in cotton) increased from

0.7% in 2003 to around 6% in 2004. In China and India, where farmers work under harsh conditions and often suffer from the effects of their own pesticide spraying, biotech crops are argued to be beneficial. The Seeds Policy, 2001 has claimed that Bt. cotton can significantly increase crop yields. With GM cotton, the consumption of pesticides can also reduce. However, there is also a cautionary message in the Policy that there is a need to adhere to safety norms like environmental, health and biodiversity safety before commercial release.

Pesticide Resistance in Agriculture

Pesticide resistance in agriculture was first noticed in India in 1963 when a number of serious pests were reported to have become resistant to DDT and HCH (two of the most commonly used pesticides during the 1960s and 1970s). Since then the number of pests with pesticide resistance has increased. During 1984-86, the excessive and indiscriminate use of pesticides on cotton created another undesirable situation viz., development of resistance in Helicoverpa armigera. The recent experience in Punjab, Rajasthan and Haryana during 2000-01 crop season is a repeat story of Helicoverpa armigera damage on cotton. In addition, wide spread resurgence of whitefly in cotton in the states of Andhra Pradesh, Gujarat, Karnataka, Tamil Nadu and Maharashtra has also been reported in the past due to pesticide use and misuse. The most serious problem of resistance is witnessed in cotton, for which American bollworm is a serious pest. The bollworm has developed resistance to almost all pesticides in a number of regions, and is particularly serious in parts of Punjab, Haryana, Andhra Pradesh, Karnataka and Maharashtra.

Growing pesticide resistance has meant that a large proportion of agricultural production is lost to pests. According to some estimates, these losses amount to between 20-30% of total production. The losses are particularly serious in cotton. For example, cotton production in Punjab declined by about 50% during 1997 and 1998, causing a number of cotton farmers to commit suicide in the affected areas. Pesticide resistance has mainly been caused by excessive and indiscriminate use of pesticides. Pesticides of spurious quality, which are commonly sold in small towns and villages, have also contributed to resistance in many areas. For example, in Bidar (Karnataka) where the problem of pest resistance became extremely serious, more than 50 brands of pesticides were found to be sub-standard in 1998-99. In another incident, the licenses of 115 pesticide producers were cancelled in

Punjab for sub-standard pesticide. Sub-standard pesticides contribute to resistance as the pests are repeatedly exposed to a low concentration of pesticides. This contributes to the build-up of resistance, without destroying the pests.

Long-Term Lower Usage Caused by IPM

Pest management is one major area of concern for sustainable development in agriculture. While pesticides enhance crop protection, a study by the International Food Production Research Institute (IFPRI) has also highlighted the paradox between the increase of global crop losses over time and the growth of chemical pesticide use. If current trends continue, dependence solely on chemical pesticides will not be a sustainable solution, from either an economic or an environmental point of view.

During the 1960s, due to problems with pesticide resistance in insect populations, a backlash against the agricultural pesticide revolution took place. As a consequence, Integrated Pest Management (IPM) was developed by entomologists as a technical system to minimise the occurrence of pesticide resistance, and to sustain the long-term effectiveness of pest control. A second major factor that boosted interest in IPM was the concern about environmental impacts of pesticides, particularly the group of organochlorines. In the early 1970s, DDT was banned in many industrialised countries. The 2001 Stockholm Convention on Persistent Organic Pollutants (POP) enacted a global ban of DDT, except for limited use in public health. Another important achievement was the 1998 agreement by ministers and representatives from 57 countries on an International Legally Binding Instrument for the Application of the Prior Informed Consent Procedure for Certain Hazardous Chemicals and Pesticides in International Trade (also known as the Rotterdam Convention on Hazardous Chemicals and Pesticides). The Convention requires that hazardous chemicals and pesticides that have been banned or severely restricted in at least two countries shall not be exported unless explicitly agreed by the importing country. The Convention enables importing countries to decide which potentially hazardous chemicals they want to receive and to exclude those they cannot manage safely. Most of the Parties of the Rotterdam Convention, so far, are developing countries.

IPM is now seen as the way forward to achieve sustainable agricultural production with lesser pesticide usage and lesser damage to the environment.

IPM is commonly referred to as the coordinated use of pest and environmental information to manage pests and keep them below damaging levels, using control options that range from cultural practices to chemicals. In practice, IPM ranges from chemically based systems that involve the targeted and judicious use of synthetic pesticides, to biologically intensive approaches that manage pests primarily or fully through non-chemical means. IPM is an ecological approach that gives highest priority to the prevention of pest problems, thereby reducing the need for pesticides. This is done through the optimal use of natural resources (e.g. maintaining healthy soil); the elimination of all farm operations that have a negative impact on the agro-ecosystem (e.g. inappropriate or insufficient crop rotation); and the protection and augmentation of natural enemies of crop pests. Next, IPM uses monitoring and forecasting to determine when pests have exceeded an 'economic threshold'. Finally, if pests exceed the threshold, pesticides may be used, with preference given to pesticides that will cause least harm to beneficial organisms and the agro-ecosystem. Thus, under IPM, pesticides are not the first recourse as a primary tool for pest control. They are used only when prevention has failed. The rationale for IPM comes from the limitations in the long-term effectiveness of synthetic chemicals for pest management, and the concerns about the negative health and environmental side effects of their use.

IPM is more an approach to sustainable crop protection than a technology or a technology package. In IPM, the available techniques are combined in an integrated management strategy aimed at keeping pest levels below a desirable level. Often, the level is determined by the damage in economic terms.

In India, traditional agriculture practiced in Indian villages till the 1960s were in harmony with nature. It preserved the diverse life forms with the agro-ecosystems and promoted the conservation of bio-diversity within farming systems. However, in the wake of the Green Revolution, several major and minor irrigation projects in India improved the water availability for cultivation of crops. The introduction of high yielding varieties in vast areas necessitated increased use of pesticides and fertilizers. The higher dose of pesticides coupled with repeated uses increased the production cost of cotton cultivation manifold. Combined with lower cotton prices in the late-1990s, many farmers could not repay their loans. Unable to repay the loans raised, a large number of cotton farmers in Andhra Pradesh,

Karnataka and Punjab committed suicide during 2000-02. The area under cotton cultivation dwindled rapidly in most of traditional cotton growing states.

In this context, the total likely benefits of IPM could be significant in terms of lower production costs. IPM can be successfully adopted in most of the agricultural and horticultural crops by optimally utilising locally available resources and farmers can save expenditure on pesticides, including the ones that are imported. Foreign exchange resources thus saved can be used for other productive programmes.

Technology Trends in the Industry

The role of PAC in accelerating agricultural production has come under scrutiny as the public has learned more about the impact of unsafe production methods and persistent organic pollutants on communities and the environment. As a result, the PAC industry increasingly needs a multifaceted approach to solve its problems: introduction of clean technologies, revamping old plants, treating effluents, establishing safety standards and guidelines, as well as an ongoing promotion and formulation of user- and environment-friendly pesticides. Responding to the economic, social and environmental pressures and more stringent requirements, there have been significant developments in PAC products and technologies. Starting from Dithiocarbamates in 1930, and DDT in 1942, PAC have evolved into organosphosphates, carbamates, synthetic pyrethroids, and later into still new discoveries.

At present, there are three major directions in research effort of PAC companies in crop protection product development:

- Many of the major multinational companies (MNCs) are replacing development of pesticides by genetic engineering to develop insecticidal and/or resistant traits.
- Companies are continuing to expand their product portfolio by investing in the research of new chemical compounds, mainly those that have innovative modes of action.
- In response to increasing levels of regulation in the developed economies, the PAC industry is aiming to develop environmentally friendly and

sustainable crop protection products that are more specific and targeted, less toxic, less harmful to beneficial organisms, and effective while using smaller amounts of active ingredients.

Product development also is focusing on preventing undesirable side effects of existing chemicals, by improving safety, reducing toxicity, and minimisation of environmental damage. For example, specialised application devices can reduce the amount of pesticides used as well as the risks for human health and the environment. In addition, appropriate package sizes are being introduced in developing countries for small-scale farmers. The industry strives to find threshold levels for optimal product use and the best practice in combining and handling pesticides.

In India, there have been significant efforts by the Government Research Laboratories, PAC industry, and several other agencies in the field of research & development (R&D), studies for bio-efficacy, toxicology, environmental impact as well as technology commercialisation. The R&D activities of the research laboratories and the PAC have yielded significant results. Advanced processes have been developed for several important products including Monocrotophos, Neem Extracts, Butachlor, Anilphos, Edifenphos, pheromones etc. The development and manufacture of new pesticides is also important not only for reducing the dependence on continued imports and the foreign exchange outgo but also for increased exports.

The GoI is also encouraging research on biopesticides. The laboratories under the Council of Scientific & Industrial Research (CSIR) have developed processes for production of biopesticides from neem seeds and custard apple. Neem-based pesticides have a broad insect pest spectrum but show low mammalian toxicity. They can be locally produced and ground seeds need only to be diluted in water to produce an active solution.

Though there are several constraints in discovering new products in India, it is not beyond India's capabilities to develop new processes for the internationally proven and well established products with improved characteristics including low toxicity, high specificity and suitability to Indian conditions. PAC companies have a dual responsibility of finding innovative solutions to the pest problems in key crops like cotton, wheat, rice, sugarcane, vegetables, etc. with their existing product portfolio. They also need to conduct R&D to conduct trials and register new molecules to offer the latest technology solutions to farmers to tackle difficult problems in controlling pests in the local agro-climatic conditions.

The R&D facilities of the Indian PAC industry have focussed on process development and on new pesticides by way of reverse engineering. The industry has not devoted enough resources to developing new molecules. The new patent protection has commenced from 2005, which will grant patent protection for 20 years for products that have been patented in the country after 1995. Most of the products in the agrochemical industry will not be covered under the patent regime as they have been patented before 1995.

Lower Pesticide Usage Potential because of Genetically Modified (GM) Crops

Over the last few decades, biotechnology research in plant sciences has made significant advances. In a broad sense, biotechnology includes techniques such as host plant resistance breeding, molecular markers, and insect population control. Transgenic techniques have the potential to reduce pesticide usage and increase productivity through introducing pest resistance traits in high-yielding varieties and to combat pathogens where no adequate control measures existed before.

Biotechnology advocates believe that GM/transgenic crops can increase food production, significantly reduce the use of pesticides, and even create drought-resistant crops that can grow on land now regarded as non-arable. The use of GM crop protection applications as alternates for some conventional pesticide applications, using herbicide tolerance and/or Bt. genes in soyabean corn, cotton, and canola have resulted in substantial savings in conventional pesticides. In the US in 2001, herbicide tolerant and Bt. crops reduced conventional pesticide use by 20.7 million kgs. of active ingredient. Similarly, in China in 2001, insecticide application on cotton was reduced by 78,000 tons of formulated product due to the deployment of 1.5 million hectares of Bt. cotton. The potential global saving of insecticides through optimising deployment of Bt. cotton alone is estimated at 33,000 MT of the 81,200 MT, applied globally on cotton in 2001.

Since the introduction of GM crops in 1996, their area under production has grown from 1.7 million ha in 1996 to 67 million ha in 2003, and to 81 million ha in 2004. In 2004, the global area of GM/transgenic crops increased at 20% compared with 15% in 2003, and 12% in 2002. The estimated global area of GM crops for 2004 was 81 million hectares (200 million acres) by 8.25 million farmers in 17 countries, an increase from 7 million farmers in 18 countries in 2003. The area increased by 13.3 million hectares or 33 million acres during 2004. In 2004, six principal countries, compared with six in 2003, grew 99% of the global GM crop area: US (47.6 million ha); Argentina (16.2), Canada (5.4), Brazil (5.0), China (3.7), and Paraguay (1.2). Canada (4.4), Brazil (3), China (2.8), and South Africa (0.4). Based on annual percentage growth in area, of the eight leading biotech crop countries, India had the highest percentage year-on-year growth in 2004 with an increase of 400% in Bt. cotton area over 2003, followed by Uruguay (200%), Australia (100%), Brazil (66%), and China (32%). In 2004, India increased its area of approved Bt. cotton, introduced only two years ago, from approximately 100,000 hectares in 2003 to 500,000 hectares in 2004. Notably, India planted 45,000 hectares of commercial Bt. cotton for the first time in 2002.

Globally, in 2004, growth continued in all four commercialised GM crops: GM soybean occupied 48.4 million hectares in 2004, up from 41.4 million hectares in 2003; GM maize was planted on 19.3 million hectares, compared with 15.5 million hectares in 2003; GM/transgenic cotton was grown on 9 million hectares, compared with 7.2 million hectares in 2002; and GM canola occupied 4.3 million hectares, up from 3.6 million hectares in 2002. In terms of adoption, 56% of the 86 million hectares of soybean planted globally during 2004 were GM/transgenic, as compared with 55% in 2003. An estimated 28% of the 32 million hectares of cotton were GM, up from 21% in 2003. The area planted to GM/transgenic canola also increased from 16% in 2002 to 19% in 2004. Finally, of the 140 million hectares of maize grown globally, 14% was GM in 2004, compared with 11% in 2003. If the global areas (conventional and biotech) of these four principal biotech crops are aggregated, the total area is 284 million hectares of which 29% was GM in 2004, up from 25% in 2003. In cotton, China increased its Bt. cotton area for the sixth consecutive year from 2.8 million hectares in 2003 to 3.7 million hectares in 2004, equivalent to 66% of the total cotton area of 5.6 million hectares in 2004. Because of their economics, of improved production the introduction of GM crops has forced increased competition in herbicide and insecticide markets. Prices of many herbicides and insecticides have declined in order to compete with the improved economics of biotechnology seed/chemical solutions. Such price reductions have led to significant discounting of weed and insect control programs and have benefited even farmers who have not adopted biotechnology crops.

World Production and Adoption of GM Crops

	Area under GM Crop (million ha)			Area under GM Crop as percent of Total Crop Area		
	2002	2003	2004	2002	2003	2004
GM soyabean	36.5	41.4	48.4	51%	55%	56%
GM maize	12.4	15.5	19.3	9%	11%	14%
Transgenic cotton	6.8	7.2	9.0	20%	21%	28%
GM canola	3.0	3.6	4.3	12%	16%	19%
Total	58.7	67.7	81.0	22%	25%	29%

Compiled by INGRES.

In 2004, the global market value of biotech crops was estimated at US$4.70 billion. The market value of the global biotech crop market is based on the sale price of biotech seed plus any technology fees that apply. The accumulated global value for the nine-year period (1996 to 2004), since biotech crops were first commercialized in 1996, is US$24 billion. The global value of the biotech crop market is projected at more than US$5 billion for 2005.

Despite the ongoing debate about GM crops, the global area and the number of farmers planting GM crops is likely to grow to approximately 150 million ha by 2010, with upto 15 million farmers growing GM crops in 30 or more, countries. Established GM country markets are continuing to grow in GM area, with a more diversified portfolio of GM crop products available. New GM countries like India and Brazil have increased their hectarage of Bt. cotton and herbicide-tolerant soybean respectively.

Bt. cotton contains the gene for Cry1Ac, which provides a fairly high degree of resistance to the American bollworm, the spotted bollworm, and the pink bollworm, all of which are major insect pests in India. Under Indian conditions, bollworms have a high destructive capacity that is not well controlled in conventional cotton The technology was developed by Monsanto, US and was introduced into several Indian hybrids in collaboration with the Maharashtra Hybrid Seed Company (MAHYCO). The first contained field trials with Bt. hybrids in India were conducted in 1997. In subsequent years, field tests were extended to collect agronomic data and information for bio- and food-safety evaluation. In 2002, Bt. cotton technology was commercially approved, and farmers have started to adopt the new hybrids. In 2001, field trials were carried out on 395 farms in seven states of India. These trials were initiated by MAHYCO and supervised by regulatory authorities. Although the sites were visited by agronomists in regular intervals for pest scouting and data collection, the trials were managed by the farmers themselves using customary practices. Insecticide amounts on Bt. plots were reduced by almost 70%, both in terms of commercial products and active ingredients. Most of these reductions occurred in highly hazardous chemicals, such as organophosphates, carbamates, and synthetic pyrethroids. In financial terms, the pesticide savings were worth about US$30 per ha. Yet the more sizeable benefits are due to yield advantages. Average yields of Bt. hybrids exceeded those of non-Bt. counterparts by 80%. Over the 4-year period from 1998 to 2001, Bt. hybrids showed an average advantage of 60%. Indian farmers would have to triple their current pesticide use in conventional cotton in order to achieve a level of damage control similar to that provided by Bt. technology.

In India, Although the field trials were managed by farmers, it is possible that average technology gains could be somewhat lower in commercial agriculture. Still, given the magnitude, the yield effects of Bt. cotton in India should remain sizeable. Many developing countries are currently in the process of assessing the costs and benefits of importing GM crop technologies from abroad for adaptation and use in their domestic agricultural sectors. Pest pressure and related crop damage vary greatly from region to region and even across individual locations. Generally, however, pest pressure in developed countries and other temperate zones is moderate, whereas in tropical and subtropical regions it is high. On the basis of pest pressure and current crop protection, the biggest yield gains from GM crops are expected in South and Southeast Asia and Sub-saharan Africa. The field-trial results from India and evidence from Indonesia and South Africa are in line with this hypothesis. South and Southeast Asia and Sub-saharan Africa are also the regions with highest population growth, so increases in agricultural output per unit area are vital for poverty alleviation and food security.

Excessive Pesticide Residues in Food, and Increased Demand for Food Safety and Quality

Acute food contamination through the use of pesticides is an issue above all in the developing countries. According to estimates, such contamination arises 13 times more often in such developing countries as in industrial countries. The reason for this is often that pesticides are used indiscriminately and the security instructions and necessary protection measures are not respected. This arises as a result of insufficient training and provision of information.

In India, excessive and inappropriate use of chemical pesticides results in the presence of pesticide residue in food. For example, according to one study, more than 80% of milk samples tested in India were found to contain residues of DDT and HCH. According to another study, residue of DDT and benzene hexachloride, both suspected carcinogens, were found in breast milk samples collected from mothers in Punjab. The amount of residue was very high and babies were ingesting 21 times the amount of these chemicals considered acceptable through their mothers' milk. Pesticide residues have often been detected in most part of the country in food grains, vegetables, fruits, oils, feed and fodder, fibres, etc. About 72% of the food samples show presence of pesticide residues within tolerance level and in 25% samples they are above the tolerance levels. BHC and DDT residue have been detected in the milk samples including human milk. Similarly, the residue of these two pesticides has also been found in egg and spices. Residues of other pesticides like organophosphorus, carbamates, synthetic pyrethroids, organochlorine pesticides like Aldrin, Dieldrin, Chlordane, Lindane, Endosulphan etc. have also been found. This indicates the seriousness of pesticide residue hazards prevailing in the country.

In India, while the Registration Committee (RC) registers pesticides for their usage, their Maximum Residue Levels (MRL) in food and commodities are prescribed by the Ministry of Health and Family Welfare (MoHFW) under Prevention of Food Adulteration Act, 1954 (PFA). The MRL is established taking into account the toxicological data of the pesticide as well as the trials on crops under good agricultural practices. At present, tolerance limits for 71 pesticides have already been notified under PFA. For 50 pesticides, tolerance limits have been finalised and the draft notification has been issued by the MoHFW. In addition, there are 27 pesticides which do not require fixation of tolerance limits. Of the balance, tolerance limits are yet to be fixed. Many of the pesticides currently being used have a tendency to survive in plants for a long time. They also enter the food chain and are found in meat and dairy products. The problem of pesticide residue is already a serious threat to health and environment in India. The incidence of pesticide residue is much higher in India than in developed countries.

With increased consumer awareness about the negative impact of pesticides on human health, consumers increasingly demand that environmental and social considerations in agriculture be taken into account. Increasingly influenced by advocacy Non-Governmental Organisations (NGOs) and actual or perceived food safety problems, consumers are showing their preferences for food safety and environmental and social values in the marketplace.

The rapid growth of organic products has enhanced the public's awareness that there are alternatives to conventional food products, containing pesticide residues. Regulations directly influence the pesticide use on produce destined for these important markets. Quality standards for export crops are increasing. Developing countries repeatedly have found that shipments with pesticide residues surpassing the limits have been rejected. This adherence to standards has driven some of the smaller export companies out of business.

Food processors and retailers are also setting higher standards regarding product quality. The food industry is looking for ways to ensure the high quality of their products, for example, concerning levels of pesticide residues, with demands for traceability of commodities, leading to an increasing role for certification, quality assurance and product labeling. Originally a phenomenon observed in industrialised countries, this trend is expanding to many developing countries. Domestic consumers, especially from the well-educated urban middle class, follow global consumption patterns and are demanding higher food safety standards.

Global Market Trends for Pest Control Products

The markets for chemical pesticides have undergone rapid changes over the last decade. Overall pesticide use worldwide over the last decade has remained constant or declined. In 2003, the value of the global market for conventional chemical crop protection products or pesticides (excluding the sales of agricultural biotechnology products) increased by 6.2% to reach US$26,710 million. The world market has expanded from US$23,170 million in 1990. Throughout this period growth has not been uniform, with the nominal value of the market in US dollar terms declining in each of the last six years.

The PAC industries in developing countries are mostly based on simple non-proprietary products. In an attempt to avoid fierce price competition from developing countries, many PAC companies in industrialised countries are switching production from basic PACs to high value-added products. Reflecting this shift, the PAC industries of developed countries are based increasingly on technical knowledge, abundant capital, and skilled management and labour.

The PAC industry of the industrialised countries has witnessed consolidation through acquisitions and mergers resulting in a few global research-based companies. The 27 large and medium-sized research-based agrochemical companies that existed in 1983 declined to 8 in 2003—Syngenta, Monsanto, DuPont, BASF, Bayer, Dow AgroSciences, Sumitomo, and FMC. The first six largest multinational agrochemical corporations account for about 85% of the total annual worldwide pesticide sales.

Several forces have driven the consolidation of the pesticide industry. R&D and registration of new products are becoming increasingly costly. The European Crop Protection Association has estimated a total cost of US$200 million for the development and bringing to market of a new chemical crop protection product.

With such high costs, there is a clear need for a regulatory process that is consistent and predictable. Further, some companies have adopted the concept of life science (pharmaceutical-agrochemical-seeds) in the company profile. This orientation contributes to a broader perspective over the entire plant production system, because it integrates pesticide and seed technology development. Globalised trade patterns also necessitate a company's presence in all major markets to ensure rapid market penetration with new products and fast recovery of R&D investments. Research-based MNCs focus on product development for crops with a large area in markets in Japan, North America, and Western Europe. They consider other markets, such as minor crops in high-value markets of developed countries and many crops in developing countries, to be of secondary importance. Because niche markets do not offer enough economic incentives for large corporations to invest in product development, fewer and fewer innovative pesticide products are being made available to the small markets for crop production in developing countries.

Intuitively, one might expect that IPM would benefit from this situation, because farmers' attention would shift to non-chemical tools. However, the opposite has happened. Farmers and extension services are still caught up in the thinking of the Green Revolution, relying on pesticides as quick solutions for their pest management problems. The pesticide industry is investing large amounts to advertise their products worldwide. In addition, the increasing availability of generic, non-brand, off-patent pesticides, which tend to be inexpensive but more toxic for human health and the environment, is not benefiting IPM adoption. The spread of generics is a disincentive to IPM systems since those pesticides, especially insecticides, tend to be broad-spectrum compounds that destroy the agro-ecosystem balance by killing natural predators that help to keep pests below economically damaging levels. Since generic pesticides are not produced under a respected company's brand name, they involve variable safety regulations. A recent study from Pakistan showed that an increase in the share of generics in the pesticide market has contributed to the deterioration of standards in quality assurance, distribution, use and disposal. In some developing countries, such as China and India, the production and use of generic compounds is increasing.

Trends in Production, Consumption, Price, Capacity Utilisation

Pesticides Installed Capacity and Production

(Thousand tonnes)

FY	2000	2001	2002	2003	2004
Installed Capacity					
Insecticides	96.60	NA	103.35	99.90	112.1
Fungicides	12.50	NA	14.09	16.10	16.0
Herbicides/Weedicides	16.14	NA	16.59	18.54	13.2
Others	3.65	NA	3.61	3.65	3.7
Total	**128.89**		**137.64**	**138.19**	**145.0**
Production					
Insecticides	70.20	69.70	59.44	49.89	49.4
Fungicides	12.60	12.09	13.59	12.27	12.1
Herbicides/Weedicides	10.00	7.25	6.10	3.44	4.9
Others	2.54	3.22	2.67	2.37	1.6
Total	**95.33**	**92.27**	**81.80**	**67.97**	**68.0**

Product-wise Installed Capacity, Production, and Capacity Utilisation

	Installed Capacity ('000 MT) at end-March		Production ('000 MT)					Capacity Utilisation	
	2003	2004	2000	2001	2002	2003	2004	2003	2004
INSECTICIDES									
DDT	6.3	6.3	3.6	3.8	3.5	2.9	3.9	46.6%	61.9%
Malathion	9.4	13.1	6.0	5.9	5.6	4.2	4.0	45.2%	30.5%
Parathion (Methyl)	4.0	4.0	1.9	2.0	2.1	1.9	1.6	47.5%	40.0%
Dimethoate	0.8	3.2	1.4	1.5	0.8	0.8	1.1	94.8%	34.4%
DDVP	3.9	4.3	2.5	2.6	2.8	2.4	2.9	62.2%	67.4%
Quinalphos	5.6	5.3	2.2	2.6	2.1	1.8	2.4	31.6%	45.3%
Monocrotophos	16.1	16.7	9.5	8.3	6.7	6.5	7.4	40.5%	44.3%
Phospamidon	5.7	6.5	4.7	3.5	0.5	0.8	0.5	14.6%	7.7%
Phorate	7.5	10.9	6.1	6.1	4.8	3.2	3.4	42.1%	31.2%
Ethion	5.1	2.2	3.4	3.5	4.1	1.2	2.2	24.4%	100.0%

Contd...

Contd...									
Endosulphan	10.1	10.1	8.3	8.5	4.5	3.7	3.2	36.3%	31.7%
Fenvalarate	2.1	1.9	1.4	1.6	1.2	0.5	0.8	24.9%	42.1%
Cypermethrin	4.6	5.3	3.8	4.4	5.1	5.1	5.4	110.4%	101.9%
Anilophos	0.6	1.2	0.9	0.8	0.6	0.4	0.3	59.0%	25.0%
Acephate	4.8	7.5	2.9	3.1	4.4	4.8	3.9	100.8%	52.0%
Chlorpyriphos	10.3	10.5	7.5	8.0	7.0	6.3	4.5	61.3%	42.9%
Phosalone	1.0	1.0	0.5	0.6	0.5	0.4	0.1	43.8%	10.0%
Metasystox			0.7	0.6	0.7	0.5	0.6		
Abate			0.2	0.3	0.0	0.0	0.0		
Fenthion			0.2	0.2	0.1	0.4	0.2		
Trizaphos			0.8	0.8	1.5	1.2	0.3		
Lindane	1.3	1.2	1.1	0.5	0.3	0.3	0.2	25.5%	16.7%
Temphos	0.1	0.2	0.0	0.2	0.2	0.1	0.1	122.0%	50.0%
Deltamethrin	0.2	0.3	0.1	0.1	0.1	0.2	0.2	92.0%	66.7%
Alphamethrin	0.4	0.4	0.4	0.1	0.3	0.2	0.2	48.5%	50.0%
Total	**99.9**	**112.1**	**70.2**	**69.7**	**59.4**	**49.9**	**49.4**	**49.9%**	**44.1%**
FUNGICIDES									
Captain & Captafol	1.8	1.8	1.1	1.4	1.2	0.8	0.6	43.4%	33.3%
Carbandazim	1.2	1.5	0.9	0.7	0.7	1.3	1.1	105.3%	73.3%
Copperoxychloride	1.5	1.5	0.2		0.0		0.2		13.3%
Ziram	0.4			0.1					
Calixn	0.2	0.2	0.0	0.0	0.1	0.0	0.0	16.5%	5.0%
Mancozeb	11.0	11.0	10.3	9.9	11.6	10.2	10.2	92.6%	92.7%
Total	**16.1**	**16.0**	**12.6**	**12.1**	**13.6**	**12.3**	**12.1**	**76.2%**	**75.7%**
HERBICIDES									
2,4-D	2.9	2.3	1.3	1.3	0.2	0.0	0.2	0.0%	8.7%
Butachlor	0.9	0.9	0.7	0.2	0.4	0.2	0.3	27.1%	33.3%
Total	**3.8**	**3.2**	**2.1**	**1.5**	**0.6**	**0.2**	**0.5**	**6.4%**	**15.6%**
WEEDICIDES									
Isoproturon	8.5	6.4	4.6	3.8	3.8	2.7	4.1	31.3%	64.1%
Basalin									
Glyphosphate	1.8	3.2	1.7	0.7	0.4	0.1	0.2	5.9%	6.3%
									Contd...

Contd...									
Paraquat	4.0		1.4	1.2	1.0	0.0		0.0%	
Diuron	0.1	0.1	0.0	0.0	0.0	0.0	0.1	48.0%	50.0%
Afrazin	0.0	0.0	0.1	0.0	0.2	0.2	0.0	500.0%	75.0%
Fluchlorine	0.3	0.3	0.2	0.1	0.1	0.2	0.0	61.7%	6.7%
Total	**14.7**	**10.0**	**7.9**	**5.7**	**5.5**	**3.2**	**4.4**	**21.7%**	**43.8%**
RODENTICIDES									
Zinc Phosphide	0.9	0.9	0.5	0.6	0.3	0.2	0.2	26.1%	22.2%
Total	**0.9**	**0.9**	**0.5**	**0.6**	**0.3**	**0.2**	**0.2**	**26.1%**	**22.2%**
FUMIGANTS									
Aluminium Phosphide	2.3	2.3	1.8	2.5	2.2	2.0	1.3	86.6%	56.5%
Methyle bromide	0.3	0.3	0.1	0.1	0.0	0.1	0.0	18.7%	0.0%
Dicofol	0.2	0.2	0.1	0.1	0.1	0.1	0.1	60.0%	53.3%
Total	**2.8**	**2.8**	**2.1**	**2.6**	**2.3**	**2.1**	**1.4**	**77.7%**	**50.2%**
All Pesticides-Total	**138.2**	**145.0**	**95.3**	**92.3**	**81.8**	**68.0**	**68.0**	**49.2%**	**46.9%**

State-wise Consumption of Pesticides

MT

FY	1993	1994	1995	1996	1997	1998	1999	2000	2001	2002	5-year CAGR
AP	13,650	10,805	9,343	10,957	8,702	7,298	4,741	4,054	4,000	3,850	-15.0%
Bihar	1,423	1,492	1,462	1,383	1,039	1,150	834	832	853	890	-3.0%
Gujarat	4,792	4,531	4,985	4,560	4,545	4,642	4,803	3,646	2,822	4,100	-2.0%
Haryana	5,200	5,220	5,100	5,100	5,040	5,045	5,035	5,025	5,025	5,020	-0.1%
Karnataka	4,125	4,164	3,640	3,924	3,665	2,962	2,600	2,484	2,020	2,500	-7.4%
Kerala	700	682	1,384	1,280	1,141	602	1,161	1,069	754	1,345	3.3%
MP	2,134	2,432	2,771	1,748	1,159	1,641	1,643	1,528	871	714	-9.2%
Maharashtra	6,450	3,021	3,647	5,097	4,567	3,649	3,468	3,614	3,239	3,135	-7.2%
Orissa	2,100	1,096	1,580	1,293	885	924	942	998	1,006	1,018	2.8%
Punjab	6,300	7,400	7,300	7,200	7,300	7,150	6,760	6,972	7,005	7,200	-0.3%
Rajasthan	3,200	3,266	3,308	3,210	3,075	3,211	3,465	2,547	3,040	4,628	8.5%
Tamil Nadu	5,500	5,177	3,394	2,080	1,851	1,809	1,730	1,685	1,668	1,576	-3.2%
UP	9,000	8,254	7,970	8,110	7,959	7,444	7,419	7,459	7,023	6,951	-2.7%
West Bengal	4,625	4,785	4,370	4,213	4,291	3,882	3,678	3,370	3,250	3,180	-5.8%
Others	1,595	1,326	1,103	1,105	895	830	878	912	1,008	913	0.4%
Total	70,794	63,651	61,357	61,260	56,114	52,239	49,157	46,195	43,584	47,020	-3.5%

Crop-wise Consumption of Pesticides in India

Crop	Share of Pesticide Use
Cotton	44.5%
Paddy	22.8%
Jowar	8.9%
Fruits and vegetables	7.0%
Wheat	6.4%
Arhar	2.8%
Other	7.6%
Total	**100.0%**

India's Pesticides Exports

FY	2000	2001	2002	2003	2004	2005 (9M)
Value – Rs.million						
Insecticides	8,942	10,853	12,197	13,305	14,562	10,751
Fungicides	463	684	722	814	1,362	1,558
Herbicides/Weedicides	276	443	381	436	424	383
Others	370	173	264	320	1,107	718
Total	**10,050**	**12,153**	**13,565**	**14,875**	**17,456**	**13,410**
Volume – MT						
Insecticides	31,156	36,162	37,816	42,978	49,072	35,255
Fungicides	4,451	6,536	7,515	7,682	12,083	10,568
Herbicides/Weedicides	2,598	3,234	2,237	2,093	3,032	2,641
Others	2,446	1,997	2,517	2,940	4,794	3,601
Total	**40,651**	**47,929**	**50,085**	**55,693**	**68,981**	**52,066**

India's Pesticides Imports

FY	2000	2001	2002	2003	2004	2005 (9M)
Value – Rs.million						
Insecticides	1,772	1,461	2,375	1,626	3,106	3,249
Fungicides	158	89	114	103	217	333
Herbicides/Weedicides	338	553	636	591	671	631
Others	101	107	494	549	1,017	1,381
Total	**2,369**	**2,210**	**3,620**	**2,870**	**5,011**	**5,594**

Contd...

Contd...						
Volume – MT						
Insecticides	3,117	2,425	4,151	3,415	7,807	8,463
Fungicides	259	223	755	674	1,138	1,369
Herbicides/Weedicides	1,606	2,919	1,677	2,138	1,681	1,821
Others	398	368	657	569	2,022	1,584
Total	**5,379**	**5,934**	**7,240**	**6,795**	**12,647**	**13,236**

Demand-Supply Position

The demand-supply position of pesticides for the last five years is presented below: (data is not available for FY2003 and FY2004)

Domestic Demand and Production of Pesticides

Thousand MT

FY	2000		2001		2002		2003		2004	
	DD	SS	DD	SS	DD	SS	DD	SS	DD	SS
Insecticides	26.8	70.2	28.5	69.7	28.5	59.4		49.9		
Fungicides	8.3	12.6	8.5	12.1	8.5	13.6	12.3			
Herbicides/Weedicides	7.3	10.0	7.5	7.3	7.5	6.1	3.4			
Others	1.2	2.5	2.5	3.2	2.5	2.7	2.4			
Total	**46.2**	**87.8**	**43.6**	**95.3**	**47.0**	**92.3**	**48.0**	**81.8**	**49.0**	**68.0**

Note: DD: demand, SS: supply/domestic production.

As can be seen from the table above, the demand for pesticides in India has grown at a slow pace, because of slow growth in agricultural production, decline in area under cultivation of pesticide-using agricultural products, and steady progress in the implementation of IPM. Since the 1990s, the most important factor behind the slow growth/decline in PAC consumption has been the slowdown in agricultural production. During the 1990s (1989-90 to 1999-2000), the growth of agriculture decelerated as compared to the 1980s (1979-80 to 1989-90). The overall annual growth rate of crop production declined from 3.72% to 2.29%. Productivity growth also declined from 2.99% to 1.21%. The consumption of pesticides also seems to have declined because of the propagation of the IPM approach, and increasing awareness about the hazards of pesticides.

As discussed, domestic PAC consumption is dependent on monsoons. PAC demand was adversely impacted during FY2003 because production of foodgrains and non-foodgrains was adversely affected, with the fall in kharif output sharper than that of rabi. However, demand growth was higher during FY2004 because of excess monsoons in the country, and an increase in sowing area. The encouraging factor was the high produce price for cotton, paddy and several other crops that the farmer has been able to realise during 2003. However, since the inventory of pesticides in several key markets is higher compared with normal levels, performance in the kharif season is likely to be muted.

Responding to the slow-down in domestic demand, production has also declined from 87.8 thousand MT in FY1999 to an estimated 68 thousand MT in FY2004. However, production has still been in excess of domestic demand. As a result, the Indian PAC industry has increasingly resorted to exports over the past few years. Export volumes of pesticides have increased from 40.7 thousand MT in FY2000 to 69 thousand MT in FY2004.

However, unit export realisations have declined because of the depressed state of world pesticides demand, and increased competition caused by from excess capacities.

Unit Realisation of Exports of Pesticides from India

Rs. per kg

FY	2000	2001	2002	2003	2004	2005 (9M)
Insecticides	287	300	323	310	297	305
Fungicides	104	105	96	106	113	147
Herbicides/Weedicides	106	137	170	208	140	145
Others	151	87	105	109	231	199
Total	247	254	271	267	253	258

Compiled by INGRES.

The Indian PAC industry has been protected by high (but declining) customs duties, and the GoI's policy towards self-reliance in pesticide production. Entry barriers to imports exists because of registration requirements, which can delay entry into India.

Import Growth of PAC

(Rs. million)

	FY2000	FY2001	FY2002	FY2003	FY2004	FY2005 (9M)	3-year CAGR
US	539	662	1389	1194	2071	2626	46.3%
Germany	173	241	457	565	612	395	36.4%
China	144	184	292	273	532	846	42.5%
UK	95	136	260	229	493	303	53.5%
Japan	502	409	606	175	346	528	-5.4%
France	244	164	245	138	246	186	14.4%
Others	672	414	371	296	711	710	19.8%
Total	2,369	2,210	3,620	2,870	5,011	5,594	31.4%

Compiled by INGRES.

Review of Performance

To analyse the financial performance of the Indian PAC industry, ICRA has analysed the financials of 13 players (hereinafter referred to as sample companies), who reported an Operating Income (OI) of Rs.39.35 billion during FY2004. These players include major MNCs such as Monsanto, Syngenta, and Bayer; and major Indian companies such as United Phosphorous, Excel, Rallis India, Sabero Organics, and Punjab Chemicals and Pharmaceuticals.

The financial performance of the companies in the PAC industry indicates a 30.1% increase in OI during FY2004. The increase was mainly because of increased demand for pesticides in India, caused by normal monsoon conditions. Further, although export volumes increased, price realisations were adversely impacted because of weak demand conditions in domestic and world markets.

Operating margins improved because of increased operating margins caused by lower material costs, employee costs and manufacturing expenses. Increased reliance on exports also offset the declining domestic demand. Many majors such as Monsanto, United Phosphorous, Syngenta, Bayer, and Excel Industries reported improved operating margins during FY2004. Competition in the marketplace continues to have an impact on realisations and also exerts pressure on margins. However, one positive impact of the stagnant demand is that it has forced most companies to consolidate operations, focus on cost savings, and streamline product portfolios to focus on high margin products. However, net margins were on the decline because of increased interest costs caused by higher inventories and receivables. Because of improved operating margins, the ROCE increased during FY2004.

Trends in Profitability and Returns of Indian PAC Industry

FY	2002	2003	2004
Operating Margin	4.7%	6.6%	10.3%
Raw material cost (% of cost of sales)	62.5%	70.5%	48.5%
Net Margin	3.3%	1.5%	6.8%
ROCE	14.3%	10.5%	21.2%
RONW	11.0%	4.4%	19.8%

Compiled by INGRES.

Overall, the PAC industry is expected to report improved sales during FY2005 caused by likelihood of further growth in exports. However, earnings are likely to be under pressure because of long-term trend of stagnant/declining domestic demand. Export markets present an opportunity for growth in income and profitability, and could provide a stabilising impact on revenues and margins.

Outlook

Modern techniques of farming have led to an increased dependence on chemicals in order to enhance food production and security. Pesticides are one of the major agro-inputs that have significantly contributed to food security in India. The domestic consumption of PAC is dependent on agricultural production, which is largely dependent on rainfall. However, domestic consumption of PAC has declined at a 5-year compounded annual growth rate (CAGR) of 3.5%. The overall pesticide use in India over the last decade has remained constant or declined. The reduction can be explained partly by changing crop prices, greater efficiency of pesticide use as a result of improvements in pest management practices and technology, and government policies aimed at both improving pest management practices, and in some cases targeting a reduction in pesticide use. Use of integrated nutrients through chemical fertilisers and organic sources is also being encouraged so as to improve and maintain soil fertility and make cheaper source of plant nutrients available to the crops. Apart from promoting IPM, the GoI has also been periodically reviewing the continued use/restriction/ban of specific pesticides reported to be hazardous. Because of declining domestic consumption, production has also declined in recent years. Over the years, the export market has increasingly provided a cushion to the vagaries of the domestic market. Out of the combined turnover of Rs.40 billion, the export turnover is around Rs.17.5 billion. An export-oriented strategy has the potential to deliver rapid expansion in volumes and provide better insulation from the cyclicalities of the domestic market. Over the years, Indian PAC companies have developed process technologies for many PACs, and PAC exports have increased significantly in recent years. The easy availability of raw materials, low-cost trained and skilled workforce, low overheads, and technically qualified managerial base is likely to make India an attractive sourcing destination for global MNCs. Many overseas companies are also undertaking collaborative research with local companies and institutions.

Though pesticides have contributed to India's growing food production, they have also paved the way for serious outbreaks of pests and diseases as well as damage to bio-diversity and environment. Despite use of pesticides, pest outbreaks continue in many crops; and pests like bollworms and whitefly in cotton, BPH in rice, Spodoptera litura in groundnut have developed manifold resistance due to indiscriminate use of pesticides. The ill effect of pesticides has also led to the significant natural pest-enemy population in the cotton, rice and groundnut ecosystem.

Given the current reliance on chemical pesticides, along with the uncertainties about many non-chemical approaches, it is unlikely that there will be pesticide-free agriculture over the next several decades. Rather, it is highly probable that the forms of IPM that will be encouraged will include a greater reliance on biological approaches and the judicious use of some chemical pesticides. Greater attention is likely to be paid to applying the right quantities of pesticides at the appropriate times and safeguarding the natural enemies of pests. At the same time, IPM will also incorporate the use of crop rotations and pest-resistant varieties of crops. Despite the efforts by the GoI and State Governments, the area under IPM is less than 5% of total gross cropped area. Therefore, there is an urgent need for GoI and State Governments to give major thrust for IPM with adequate budgetary support. The GoI not only need to support IPM but also regulate hazardous pesticide use.

Another important issue regarding pest management in the future centers on the role of biotechnology in crop production. The present decade is likely to see a substantial increase in the production of GM crops. Some of these crops have been engineered so that the application of herbicides will destroy weeds but not the economic crop. Other genetically engineered plants have been designed to resist pests such as stem borers and nematodes without the need for pesticides. Others are expected to combine both herbicide resistance and insect resistance in one seed. The overall effects of GM crops on pesticide use remain to be seen. It is probable that the use of herbicides may expand while the use of certain insecticides may diminish. There are concerns, however, that the rapid diffusion of improved varieties may well lead to pest resistance, including the possible transfer of genetic qualities from modified plants to weeds, creating new generations of weeds resistant to herbicides. There are also concerns about the long-term effects of increased consumption of genetically altered materials on both humans and animals.

Policymakers and stakeholders interested in effective crop protection have to balance the social benefits and costs of pesticide use. Options for improving pest management include the development of pesticides that are more benign than current products. IPM promises to be the most pragmatic approach. Genetically engineered crops offer great promise but need to be monitored. Developing countries such as India will have to invest in their own technological capacities or enter into sharing arrangements with MNCs.

Environment Analysis: Porter's Model

Threat of Substitutes: Medium
Expected to remain one of the vital ingredients of crop protection, and enhancing agricultural productivity. However, lower pesticide usage in recent years, because of environmental concerns, and Integrated Pest Management (IPM) techniques. Low short-term threat from Genetic Modified (GM) crops and organic products.

Bargaining Power of Suppliers: Low to Medium
Most of the raw materials are available easily, but prices are cyclical.

Inter-Firm Rivalry: Medium
Industry fragmented with 30-40 large manufacturers and 600 formulators. The organised sector comprises multinational corporations (MNCs) and Indian players.

Bargaining Power of Buyers: Medium
Demand for credit is high in the industry. The market is predominantly credit based with major realisations taking place immediately after the harvest season.

Barriers to Entry: Medium
Low capital requirements. Indian players focus on applied research, which include developing processes to manufacture off-patent products, more effective methods of delivering existing products, and new formulations of generic products. The MNCs dominate the market for proprietary products (patented new molecules). They have the advantage of access to the R&D pipeline and superior financial resources of their parents.

Section II

Country Perspectives

4

Your Daily Poison: The Second UK Pesticide Exposure Report

Alison Craig

The food we consume in our daily course, assuming it to be nutritious, is actually a platter of poison. We are exposed to a numerous poisonous pesticides in our meal every day without being even aware of the fact. The below mentioned report intends to enlighten the readers and the public at large, to the extent to which we are all exposed to pesticides in our food, water and the environment.

Executive Summary

This new report, the second in the series[1], examines in detail government information accessible to the public about exposures to pesticides we all have, on a daily basis, in our food, water, and the environment. The previous report covered 2003 figures, and this report covers exposures in 2004 from published sources. We highlight the paucity of accessible information available, not just to the public, but to decision-makers at a local level: local authorities, water companies and health professionals, who are responsible for protecting public health. The implications of an important new report by the Royal Commission on Environmental Pollution (RCEP) 'Crop spraying and the health of residents and bystanders'[2], for all forms of pesticide exposure, are assessed.

Our Main Findings

- An analysis of results from UK regulators and the latest European Commission data indicates that levels of pesticide residues in food are increasing in the UK and are significant across Europe, with consequent health risks.
- Survey results in this report confirm the widespread presence of pesticides as low-level contaminants in drinking water. Because of advances in technology, it has become evident that the legal limit for pesticides in treated water needs revision.
- Pesticides which have been contaminating drinking water ever since they were approved: atrazine, isoproturon, mecoprop and simazine, are still widespread in drinking water.
- The post-approvals monitoring for pesticides in food and water is inadequate, because insufficient numbers of pesticides are tested for, even though there is evidence of their use.
- Because the government has not set standards and protocols for the selection of specific pesticides to be tested for in food and water, and the analysis of results, existing strategies are inconsistent.

Introduction

This report of pesticide exposures in 2004 follows up evidence of exposures in 2003, published as the report *People's Pesticide Exposures – poisons we are exposed to every day without knowing it.* The public has the right to know the extent to which we are all exposed to pesticides in our food, water and the environment. Information in this report is sourced from government monitoring data, and set out in an accessible form alongside more detailed data we collect from the authorities by questionnaire survey, and our own results. The intention is to fill in some of the gaps in government-disclosed data, and identify areas in which information is not available either because it is not collected or because it is kept secret. Trends are also described where possible.

Since the last report, the government has published its National Strategy for the Sustainable Use of Plant Protection Products. The strategy does not encompass a plan for overall reduction of pesticide use, and it does not yet include a section on protecting human health from the effects of pesticides. The government supports an industry-controlled programme, the Voluntary Initiative, to attempt the reduction of harmful effects of pesticides on the environment, the success of which is limited[3]. Overall pesticide use remains high: over 31,000 tonnes of active ingredient are applied to UK farmland per year (Appendix 1). The government asked the Royal Commission on Environmental Pollution (RCEP) to carry out an independent investigation into the health impacts of pesticides, and its report, published in September 2005, made important recommendations to protect human health from pesticide exposure.

Sources of published government information include the Health & Safety Executive, the Pesticides Safety Directorate (Defra) the Advisory Committee on Pesticides, the National Poisons Information Service, the Pesticide Residues Committee; and the Drinking Water Inspectorate. Additional information is provided by PAN UK's own surveys.

There are summaries of the organisations involved in the regulation of pesticides and pesticide laws, and the regulatory testing and assessment of pesticides at Appendices 2 and 3.

1. Pesticide Exposure in the Environment

People's Experience of Pesticide Exposure

There are a number of monitoring schemes for the collection of pesticide exposures and poisonings: the government's Health & Safety Executive's (HSE) Pesticide Incidents Appraisal Panel (PIAP) scheme, the Pesticides Safety Directorate (PSD) annual survey of cases reported by companies (the 'human health incidents survey'), and pesticide poisoning enquiries reported by the National Health Service National Poisons Information Service. People also report poisoning incidents to their local authority. There is no aggregation of all schemes to produce an annual total of exposures reported to government. The PEX-Action on Pesticide Exposure database is run by PAN *UK*, and other non-governmental organisations, such as the Organophosphate Information Network and the UK Pesticides Campaign, also maintain records of exposure cases.

A. The Health and Safety Executive's Pesticide Incidents Appraisal Panel (PIAP)

The HSE is the government agency which has the main responsibility for the investigation of complaints alleging that pesticide exposure has caused ill-health. The HSE PIAP is comprised of a panel of government toxicologists and medical experts, and considers all incidents reported to the HSE's Field Operations Directorate where there is any allegation that the use of a pesticide has caused harm or ill health.

The Royal Commission on Environmental Pollution (RCEP), in its report 'Crop spraying and the health of residents and bystanders', confirmed PAN UK's concerns about PIAP, the main government post-approvals surveillance scheme for pesticide-related ill-health, and concluded that it is so inadequate that the HSE should relinquish this function[4]. In its evidence to the RCEP, PAN *UK* identified the following flaws:

- PIAP does not initiate clinical examination itself: the assessment largely relies on medical information from the complainant's General Practitioner (GP). But if the patient does not know which chemicals he or she has been exposed to it is impossible for the GP to investigate properly;
- GPs have very little training in toxicology, and the specialist resource available to them, the National Poisons Information Service, is for acute poisonings only;
- PIAP is designed to deal with 'incidents' only, and does not take account of disease which could be linked to chronic exposures; and
- Only after the HSE inspector has concluded his or her investigation of a pesticide complaint, and taken official actions such as issuing a notice to the farmer, does he or she send details of the alleged ill-health incident to PIAP.

This incurs a delay of, on average, six months. This means that it is very unlikely that evidence, in terms of pesticides in the environment, or in biological media, is still present. Around one percent of cases reported to PAN *UK* pursue their complaint to the stage where it is dealt with by PIAP, and none of these cases in 2004/2005 was classified as 'confirmed' or 'likely'.

Between 1 April 2004 and 31 March 2005, HSE inspectors investigated 150 pesticide related incidents, see Table 1. This was a reduction from the previous year's total of 204. Fifty five incidents involved allegations of ill health, seven less than the previous year. The HSE has acknowledged that under-reporting is a problem in respect of PIAP, and a loss of confidence in the scheme may result in fewer reports.

B. 'The Human Health Incidents Survey': Pesticide Exposures and Poisonings Reported to the PSD by Companies

The latest survey by the PSD of incidents reported by members of the public direct to pesticide companies (Appendix 4a) indicates a significant rise in cases of exposure.

Table 1: Pesticide Incidents Appraisal Panel Data for 2004/05[5]

PIAP Classification	Total		Employees/Self-Employed		Members of Public/Others	
	Incidents	People	Incidents	People	Incidents	People
Confirmed	0	0	0	0	0	0
Likely	5	8	0	0	5	8
Open assessment (i)	0	0	0	0	0	0
Open assessment (ii)	3	3	0	0	3	3
Unrelated	5	5	0	0	5	5
Insufficient	25	32	1	1	24	31
Information Pending	15	17	1	1	14	16
Not an incident	2	5	1	1	1	4
Total	55	70	3	3	52	67

Pesticide Incidents Report, HSE Field Operations Directorate Investigations, 1 April 2004-31 March 2005.

Definitions

Confirmed: There are clinical symptoms and signs typical of exposure to the cited pesticide formulation combined with either:

- corroborating medical and (where appropriate) biochemical evidence; or
- evidence of overexposure.

Likely: The balance of evidence based on reported exposure circumstances, clinical symptoms and signs or biochemical evidence (where appropriate) is consistent with ill health due to exposure to the cited pesticide formulation.

Open assessment: (i) The reported ill-health is not consistent with the known potential ill-health effects of the cited pesticide formulation given the reported exposure circumstances but the implied association cannot be entirely discounted in the light of current knowledge; or (ii) the evidence is consistent with pesticide exposure being the cause of the reported ill-health but alternative explanations, e.g. pre-existing disease are also present.

Unrelated: There is strong evidence, e.g. evidence about exposure or from medical reports, that the reported ill-health is not pesticide-related.

Insufficient information: The available data is insufficient, incomplete or conflicting and the panel is unable to classify a case for one or more of these reasons.

Reports have risen from 137 in 2002, to 177 in 2004, giving a total of 466 incidents reported by companies in three years. Cases involving children account for a steady 15 percent each year so far.

The low figures for pesticide-related ill-health reported by the HSE from the PIAP scheme have been used many times to justify reassurances by the Advisory Committee on Pesticides (ACP) to both the public and Ministers, that these substances are not a significant public health issue. The ACP is the expert committee providing advice to Ministers on pesticides (see Appendix 2). But the Human Health Incidents Survey, first published in December 2004[6], instantly doubles the official figure which for years has been estimated at 'less than 100 incidents per year' based on PIAP data.

The survey was first conducted in 2003 after PAN *UK* questioned whether or not the pesticides industry had been complying with a legal obligation to submit immediately any new information on the adverse effects of their products. PSD sent surveys to the companies but despite several extensions of the deadline, 13 out of 184 companies failed to respond. The PSD responded by revoking the licences for all products sold by the non-responding companies[7]. Revocations of 'a handful[8]' of the products were subsequently reversed on receipt of the requested information from the companies. In the most recent survey, all companies responded. As with the previous survey, the PSD has reported results in full on their website[9], but this time has withheld company names.

Examples of exposures reported by the PSD:

- Children were playing on a lawn three days after it was treated with MCPA + dichloroprop-P + dicamba + ferrous sulphate. Forty-eight hours later the children had sickness and diarrhoea.
- Two young children were playing on the lawn treated with MCPA + mecoprop-P + ferrous sulphate. Both children have developed a rash on their legs.
- A paddock was treated with a product containing clopyralid and triclopyr. Some mint was picked from it and used in cooking potatoes. A girl aged 3 who ate them was reported ill: she already had tonsillitis and a temperature.

Fatality Reported

PAN *UK* traced details of a fatality listed in the PSD results of 2004. It occurred in 2002 after exposure to aldicarb, but it was not clear whether or not it was related. The case was publicised across Scotland in the Sunday Herald newspaper[10]: the following is an extract.

An official investigation into whether a young Scottish farm worker was killed by a highly toxic pesticide had to be abandoned after a university mix-up. Graham Stephen died on May 3, 2002, after he had applied a dangerous pesticide known as aldicarb to a potato crop on a farm near Forfar in Tayside. He was 37.

After hearing from police that Stephen could have breathed in the pesticide and that this might have helped cause his death, the procurator fiscal ordered his blood to be screened for aldicarb. Samples were sent to the forensic science laboratories at the University of Glasgow, but were accidentally disposed of after six months before techniques for detecting aldicarb had been developed. The university said the incident was a 'regrettable accident'...

'Aldicarb is one of the most hazardous pesticides still licensed for use on farms in the UK. It was one of the pesticides produced at the chemical plant at Bhopal in India where an accident in which poisonous gas was released in 1984 killed 8000 people. Because of the risks aldicarb poses to human health and the environment, the European Union has severely restricted it, but a ban has been resisted in the UK and other countries, by aldicarb's multinational manufacturer, Bayer CropScience.'

PAN *UK* comment: companies are required to submit an analytical method as a condition of licensing to the regulators[11].

- A woman working as a spray operator who was three months pregnant contacted the manufacturer, Bayer CropScience company to ask if glyphosate could harm her unborn child. The company doctor closed the case, reporting 'there were no further issues'.
- A baby girl of 9 months was found by her parents lying on floor. A broken plant pot which had compost in it containing chlorpyrifos was covering her face. The child appeared to have ingested an amount of compost. She vomited three times.

C. Pesticide Poisonings Reported to the National Health Service (NHS)

The National Poisons Information Service (NPIS) is a clinical toxicology service for healthcare professionals working in the NHS and is a service commissioned by the Health Protection Agency. The service consists of a network of units across the UK, providing information and advice on the diagnosis, treatment and management of patients who may have been accidentally or deliberately poisoned. Information on management of poisoning is available to registered medical professionals on TOXBASE, an Internet database or via a 24 hour telephone service for more complex cases requiring specialist advice[12].

The NPIS changed the format of its annual reports in 2004, now reporting data in the new format for a few types of pesticide active ingredients (Appendix 4b), so comparing earlier reports directly is not possible. Telephone enquiries to the NPIS do not necessarily represent numbers of poisonings because they are often precautionary only. The NPIS deals with all kinds of poisoning: in 2004/5, the majority (67%) resulted from pharmaceuticals, and agrochemicals represented 1% of enquiries (other sources: industrial chemicals 12%, household 8%, cosmetics 8%, plants 3%).

The total number of enquiries for pesticide poisonings in 2004-05 was 739, which included 268 pyrethroids, 137 organophosphorus, 125 paraquat (including two deaths), 120 glyphosate and 89 carbamates. The top 20 pesticides reported as responsible for incidents are listed in Table 2. In the previous year (2003-04) the figures were: 102 for glyphosate, 175 for organophosphorus insecticides and 88 for paraquat with one death.

Table 2: Top 20 Agents Reported to be Involved in Exposures, April 2004 to March 2005

Ingredient	Number of Exposures
Permethrin	84
Paraquat	58
Diquat	57
Glyphosate	56
Metaldehyde	48
Bromadiolone	36
Phenols/cresols*	32
Borax	29
Sodium chlorate	23
Bendiocarb	21
1,2-benzisothiazolin-3-one	21
Difenacoum	17
Tetramethrin	14
Cypermethrin	12
Diuron	9
Chlorpyrifos	8
Pyrethrins	7
Alphachloralose	7
Organophosphate	6
Alphacypermethrin	6

* From creosote but note that creosote has been banned since 2003. However, some members of the public use 'creosote' generically to denote some sort creosote.

Source: Adams R D, Good A M, Bateman D N, 'Pesticide exposure monitoring using NPIS resources April 2004 – March 2005', NPIS Edingburgh.

The NPIS is conducting a new surveillance programme specifically for pesticides[13] using its online service TOXBASE. The survey was commissioned by the PSD and HSE following work by the ACP on improving the surveillance of pesticide-related ill-health. In total, information on 588 pesticide-related exposures was collected from telephone and TOXBASE enquiries. This represents an improvement in the surveillance and reporting of pesticide poisonings. The scheme is voluntary, and it is unknown how many more cases would be reported if it was mandatory. Although information on head lice products was also collected, it was not included in the report to the PSD, because it is the Medicines and Healthcare Products Regulatory Agency (MHRA) which regulates these products.

At their meeting in November 2005, the ACP considered the pesticide exposure monitoring report from the NPIS and commented that 'most incidents were relatively minor, and that those which were more serious tended to arise from intentional poisoning.'

D. Pesticide Exposures Reported to Local Authorities

Incidents are still reported to local authority environmental health departments despite the publication by the HSE of an accessible leaflet[14] to the public on reporting. In 2002, over half of the local authorities (see Table 3) who responded to the PAN *UK* survey had received reports of a total of 11 incidents, and in only three cases where they reported to the HSE. In 2004 six local authorities reported a total of 11 incidents, of which one was reported to the HSE. The return rate to our questionnaire was very poor; but if a half to a third of all 468 local authorities in the UK has such incidents reported to them, which never reach the HSE, this is a significant underreporting issue.

Table 3: Incidents and Bystander Exposure Reported to Local Authorities

PAN *UK* Questionnaire sent to:	Approx 468 local authorities in England, Wales, Scotland and Northern Ireland	
Number of responses received:	In 2004 (2002 data) 33	In 2005 (2004 data) 20
Number of local authority respondents reporting pesticide incidents:	In 2002 and 2003: 11	In 2004: 6
Number of incidents:	2002: 10; 2003: 6; others estimated, eg 'normally about six complaints per annum, of all kinds'	In 2004: 11

E. The Pex Database

PAN *UK*'s PEX – Action on Pesticide Exposure information service and database collects evidence of pesticide exposures and their effects on health. There are 973 entries on the database, which have accumulated since the project was initiated by Mrs. Enfys Chapman as the Pesticide Exposure Group of Sufferers (PEGS) project, passed to PAN *UK* in 1998. We also receive around 1200 enquiries per annum on pesticide exposure issues generally. As well as providing an advice and support service, the PEX project analyses these exposures, raises issues with the regulatory authorities, and campaigns for a reduction in pesticide exposure.

The routes of exposure reported to us are via inhalation and dermal exposure from spray-drift and vapour in the environment. The RCEP estimates, in the new report, that between 1 and 1.5 million people live in homes bordering agricultural and horticultural land in Great Britain, so numbers reporting to PEX are a small percentage of potential exposure cases. Illnesses caused by the ingestion of pesticides in food and water are not reported, because it is impossible to link these to health effects without prompt laboratory and biochemical tests, which are prohibitively expensive and impractical in almost all circumstances.

In 2004, we were contacted by 16 cases of new, current exposures, 12 of which were in the 'bystander' situation of living close to sprayed fields, and therefore having ongoing, repeated chronic exposures. We are aware that, in terms of numbers, only the tip of the iceberg is reported to PEX: many people exposed to pesticides do not know how to contact the right officials, or organisations such as ours. A selection of cases of people who contacted PAN *UK* in 2004-05 are described below.

Pesticide Exposure as a Possible Cause of Chronic Disease

'I have lived in this Hampshire village, of about 500 residents, for nearly thirty years. In the last ten to twelve years, I have noticed a dramatic increase in the incidence of cancer. I can think of 26 people in the village who have either been diagnosed with cancer, or who have died from it recently. Two people who died developed three primary cancers, one after another. There are numerous cases of breast cancer, but also bone cancer. I am concerned that environmental factors may be causing some of these cancers. We live adjacent to a farm of a thousand acres and crops are regularly sprayed.'

This report is consistent with a case in Gloucestershire reported previously to PEX, in which five people suffering from depression, leukaemia or cancer lived in the same row of houses backing onto sprayed fields. There are 18 examples on our database in which 'clusters' of disease have been reported to us, with the suspicion that pesticide exposure may be a contributory factor.

Residential Exposure

'I have lived in the country all my life. The farmer grows barley nearby and when they spray it fills the house. In the spring I put all my washing on the line which was then oversprayed. If they notified me this kind of thing could be avoided. I get irritable bowel syndrome when they are spraying and I am concerned for my children and being out in our garden with them. The smell from the spray can be really bad. Our garden is approximately twenty metres from field edge.'

Exposure to Spray-Drift, Pesticide Particles and Vapour

'The field next to my garden was being sprayed with a tractor and boom with Round-up, and there was a moderate wind blowing across my land. I was working in the garden and noticed the smell and taste of spray. I took photos, and went to see the farm manager, who apologised but said there was no risk: he said it was fine spray, not real 'drift'. The fruit and veg in my garden have been exposed, but the HSE will not carry out a laboratory analysis and I cannot afford to do so myself.'

Exposure to Storage Chemicals

'I and my family live in a row of houses next to fields and approximately twenty metres from a potato store regularly fogged with chlorpropham. Fans were installed without ducts or filters in, October 2001, so we are regularly exposed. No notification is given despite frequent requests. My wife has recently been diagnosed with asthma: she has never had it before. My brother in law in the same row was diagnosed with terminal cancer of the oesophagus since the fans were installed and has now sadly died. The Health & Safety Executive have inspected the site and concluded that there was no risk to our health. But in a previous case examined by the HSE's Pesticide Incident Appraisal Panel (ref 07/069, November 1995) the Panel concluded that exposure to chlorpropham was likely to have caused the illness reported because chloropham is known to have irritant properties, including respiratory tract irritation after inhalation. Now I am attempting to pursue legal action.'

Harm to Pets Indicating Risk to Children

'On 18th May 2004, Carmarthenshire County Council roadmen cut grass and sprayed chemicals around the bridges over the river Taf in the small village of Llanfallteg. That evening, as a local resident, I was walking my dog along the road, and noticed that after we'd crossed a bridge he began bumping into objects. When we arrived home, I realised he had inflamed eyes and appeared to be almost blind. The following day I took him to the vet who diagnosed chemical poisoning. I was informed by the Council's Highways Department staff that the area had been sprayed with 'Tordon22k' (herbicide picloram) and 'Touche' (herbicides diuron and glyphosate). My dog survived for a further two weeks, looking like a rabbit suffering from myxomatosis, but then had to be put down because of irreparable damage to his eyes. I am really shocked that these substances were being used with no warnings. Supposing a child had used the verge?'

Harm to Domestic Animals Indicating Risk to Human Health

'My ponies are in a field surrounded on two sides by arable land on which sugar beet is grown and sprayed frequently. One of the foals developed a swollen throat and could not put its head down to suckle properly. They were spraying about a foot away from the edge of our field. We are never notified of when they are going to spray. Our oldest pony had to be shot in April (2005), and the vet said it had neurological problems. It has only ever had such symptoms when grazed on that field.'

Exposure to Orchard Spraying

'I live next to an orchard where pears and apples are grown commercially. In the spring they spray about once every two weeks. My garden is about ten to twelve feet from the boundary. I am concerned about the use of organophosphate insecticides in orchards.'

Exposure to Fumigants Used in Polytunnel Production

'My whole community is at risk from the use of methyl bromide in a nearby strawberry-growing enterprise. Some of us have experienced background flu symptoms and there are considerable health risks from this chemical.'

Exposure to Sulphuric Acid

'I foster two people with learning disabilities who suffered burning sensations on their skin and in their eyes when sulphuric acid was used on the potato crop next to our garden near Leven. Other members of my family and our animals were also at risk. We were supposed to be notified but were not.'

PEX was also contacted in 2004 and 2005 by 32 people reporting previous pesticide exposures, for example:

- 'About ten years ago I was made really ill and was in bed for days after walking where they were spraying acid on potatoes. I don't report the spraying because I don't want to antagonise the farmer.'
- 'A few years ago I was walking down our lane with my baby in the push-chair. We were caught in spraydrift and my son started coughing. We had to rush him to hospital and he suffered severe respiratory symptoms for weeks.'
- 'I was diagnosed with breast cancer [in 2005] and have had to have a mastectomy. I do not understand why the chemical pollutants in breast tissue are not analysed and their potential role in disease investigated. I asked why I was not given a blood test for chemicals and was told 'they don't tell us much'. I want my tissue to be used in research into the underlying causes of cancer.'
- Sapperton Parish Council in Gloucestershire and Cley Parish Council in Norfolk made submissions to the PSD's public consultation in 2003, reporting that pesticide exposure of parishioners has been a regular complaint for years; PEX provided information.

Conclusions

PAN *UK* has been lobbying government since September 2002 for the new Health Protection Agency, with its broad remit to protect public health from the adverse health effects of chemicals, to institute a surveillance system with the following features:

- Accessibility with well publicised portals, e.g., GPs, NHS Direct, and the websites of all relevant government agencies including the PSD and HSE.

- Powers of prompt collection of evidence.
- Power to arrange, in conjunction with a person's GP, rapid biochemical investigation.
- Medical outcome follow-up.

The existing government schemes should be analysed and amalgamated so cases of people whose health has been affected by exposure to pesticides can be accurately assessed. It is currently not possible to estimate how many numbers occur because each scheme does not indicate whether or not cases have also been reported to the others.

2. Food Residues

PAN *UK* has conducted an analysis of residues both from the UK government's testing programme (2004 results) and from the European Union testing programme (2003, latest results), identifying the occurrence of pesticides of particular concern for human health. There are serious issues about potential health effects, particularly on the unborn, babies, and toddlers. For example, recent research suggests that exposure to carcinogens in the uterus or early childhood can damage DNA and heighten susceptibility to disease later in life[15]. Children can consume a disproportionate level of pesticide residues because of their high consumption relative to their bodyweight.

Europe

Levels of residues in food across Europe have remained high. The latest figures from the European Commission's monitoring programme indicate that conventionally produced food is as contaminated with pesticides as in 2002 (Appendix 5a). In a new report[16] the EC acknowledges that toxic chemicals in food are a risk to the health of children and vulnerable adults, reporting breaches of safety limits which are a significant risk to health. According to PAN Europe, 'These results show a complete failure of the EU in controlling the level of pesticides in our food[17].' Countries which test a relatively high number of samples, such as Germany, detect a higher number of residues. The number of samples tested per head of the population in the UK is the lowest in Europe with the exception of Portugal[18].

According to the data from the European survey, babies and toddlers are at risk from residues of some of the most acutely toxic pesticides detected in tests. At the residue levels found, a toddler could consume well over the health-based safety limit, the amount that can be safely consumed in one meal or one day, (see Acute Reference Dose (ARfD)). This would equate to 147% of chlorpyrifos in table grapes, 164% of methamidophos in sweet peppers, over twice the limit of endosulfan, over five times the limit of triazophos in sweet peppers and a staggering ten times the limit of methomyl—1035%—in table grapes.

UK

The government's Pesticide Residues Committee (PRC) runs a programme of residue testing jointly paid for by a levy on agrochemical companies and by government funding. The official source of data on pesticides used in the UK is the Central Science Laboratory's Pesticide Usage Survey Group data[19]. This is a sample of approximately 2,000 UK farms. The PRC acknowledges that 'the range of pesticides that may be used in agriculture and food production, either in this country or abroad, is very wide. About 350 active substances are currently approved for use as agricultural pesticides in the UK and over 850 are approved in one or more EU states. Potentially around 1,000 different chemicals might be sought[20].' Yet the PRC only tested for 123 pesticides in 2004[21].

Results from the UK government's testing programme (Appendix 5b) indicate rising levels of residues—25 percent of foods sampled in 2003 contained pesticide residues, and the figure rose to 31 percent in 2004. Residues below or at the legal limit (the Maximum Residue Level) contaminated 30 percent of samples (24 percent in 2003), and above the legal limit 1 percent of samples (less than 1 percent in 2003).

One sample of round lettuce was found to contain inorganic bromide at between 2.9 and 5.4 times the ARfD for children and adults. One sample of lettuce imported was found to contain two residues above the legal limit (the MRL) of endosulfan and methamidophos. A risk assessment for methamidophos estimated exposure at 9.3 times the ARfD for children. According to the PRC, 'Short-term negative effects on people's health are unlikely, but sensitive children might briefly have symptoms such as sweating, producing too much saliva or stomachs' [sic, presumably stomach ache]. One sample of imported speciality beans (yard-long beans) contained two residues above the legal limit. A risk assessment for triazophos showed that intakes for adults and infants were 1.7 and 3 times the ARfD[22].

PAN *UK* has collated the UK data and established the known hazards to health by the residues found in it. This indicates that consumers are being exposed to pesticides which are often highly toxic or are known or suspected to cause more long term health effects including the disruption of hormones (endocrine disrupting chemicals). Of the 94 pesticides detected in food according to official UK data for 2004:

- 34 have been identified by the World Health Organisation as acutely toxic;
- 37 have been identified by international authorities as suspected carcinogens;
- 22 have been identified by international authorities as suspected endocrine disrupting chemicals, implicated as possible causes of chronic disease, including cancer and reproductive disorders.

According to guidance generated by the Pesticide Residues Committee[23],

The choice of pesticides to be sought is primarily influenced by:

- Pesticide use
- Potential for residues based on use pattern and the physico-chemical properties of the pesticide
- Analytical capabilities
- Toxicological profile of the pesticide
- Existence of MRLs

A range of sources of information are used including data from previous monitoring (UK and elsewhere, Rapid Alerts), Pesticide Usage Surveys (UK only), pesticide registration data, UK and Codex MRLs and other intelligence, including contributions from [PRC] members.

> Generally the emphasis is on insecticides, fungicides and post harvest treatments since these have the greatest potential for residues. A range of approximately 80-100 pesticides can be sought using a multi-residue method. Other pesticides need to be analysed separately. Single method pesticides are therefore relatively expensive. Lower reporting limits are also more expensive.
>
> Research and Development is separately funded to develop/validate methods of analysis for pesticides new to the PRC so that these can be incorporated into the routine monitoring programme, ideally into the existing 'multi-residue method.'

This means that pesticides which can only be detected through the single method are less likely to be included in tests. Herbicides and compounds which have low reporting limits—can be detected at the lowest levels—are also less likely to be included because of their cost. Thus these pesticides could be present in our food but never tested for by government. In the approvals process tests are conducted only under controlled conditions. All newly registered substances should be included in the PRC monitoring programme because, unless they are tested for, it cannot be known if they will show up as a residue under practical conditions.

David Mason, of the Central Science Laboratory (CSL), one of the laboratories contracted by the PRC to carry out the analyses, was also a member of the PRC's Analytical Sub-group. It is the group's task to scrutinise a list of proposed analyses from PSD. Although the CSL website states 'Because we know from experience which residues we need to look for in any given foodstuff, we can offer you a package of analysis tailored to your needs', it is unlikely that the data exists to enable any agency to know accurately and comprehensively which pesticides are in foodstuffs.

Asked *how* the CSL knows which pesticides to look for in any foodstuff, David Mason said[24] 'Residue testing can never be more than spot checking. You have to strike a balance between the number of samples, the number of tests and the budget available—this is always the problem. My own view is that on a spectrum of either knowing a lot about a few samples, or much less about a lot of samples, the latter is better. We use usage data rather than what has been approved for use, because if, for example, three herbicides are put on the market one year, and all three are initially used, in time just one may be the popular one that farmers are using.'

David Mason confirmed that there could be a time-lag of a year or more between the approval of a new pesticide and its inclusion in the PRC monitoring programme. As part of the licensing agreement, companies have to submit analytical methods, but, according to David Mason, 'official laboratories including CSL develop and validate tests for the new pesticides using multiple residue methods if possible and that takes time. Now the companies have to submit, as part of the dossier for a new pesticide, an analytical method for use in multi-residue testing methods if possible. It is much easier to include a pesticide in the PRC programme if it is amenable to multi-residue testing.'

The PRC has recently disclosed in a reply to PAN *UK*[25] some 'illustrative costings' of increasing the numbers of samples tested. To increase the sample size of only 45 food items (the number currently tested) to 300 would raise the cost of the monitoring programme more than three-fold to over £7 million. The PRC considered this in 2002, but has not yet increased the numbers of samples tested.

Although the PRC 'brand name annex' publications, which list the retailer and brand name for samples tested help to drive down residues by 'naming and shaming' food retailers, the vast majority of food residue results are inaccessible to the public. Supermarkets and food suppliers have their own self-funded testing programmes, results of which are only disclosed in a limited way. Put together these exceed the government's programme in size.

Conclusions

These residue results reflect the fact that no progress has been made in reducing pesticide usage in the UK, and is urgently needed. The Food Standards Agency (FSA) has had a residue minimisation plan since 2002 which is limited to a few crops. Consumers continue to express their preference for food without residues, and multiple retailers are responding to this pressure. For example, the Co-op explicitly promotes a claim that it is 'Leading the way on pesticide reduction'. Sales of organic products continue to grow much faster than sales in the non-organic grocery market and in 2004 reached £1.213 billion, an 11 percent increase on the previous calendar year. Increasing sales of organically produced food suggest that future progress is possible.

3. Pesticides in Water

In this section we compare official data from the government's Drinking Water Inspectorate (DWI) with the results we have obtained from questionnaire survey of both the water companies who report to the DWI, summarised in Table 4 (Appendix 6a and 6b), and local authorities, summarised in Table 6. Local authorities are responsible for private water supplies, but conduct only a small number of tests. Indications of ubiquitous contamination of drinking water at low levels are confirmed through the PAN *UK* survey. A gap in the protection of health from potentially harmful pesticides in water, due to the limited number of pesticides tested for, is revealed.

Table 4: Pesticides in the Public Water Supply

Summary of PAN UK Survey – Appendix 6a

PAN *UK* Questionnaires sent to:	All 26 water companies in England and Wales, Scottish Water, and the Northern Ireland Drinking Inspectorate	
Number of responses received:	In 2004 (2002 data) 12	In 2005 (2004 data) 12
Number of companies reporting 10 or more pesticides found above 0.01 micrograms per litre (below the legal limit) in raw (untreated) water	4	5
Number of companies reporting 10 or more pesticides found above 0.01 micrograms per litre (below the legal limit) in drinking (treated) water	4	5

According to the official data from the government's Drinking Water Inspectorate, pesticides in water is a minor and diminishing cause for concern[26]. The legal limit, set by the European Commission, is 0.1 micrograms per litre, or one part in ten billion, for a single pesticide, 0.5 micrograms per litre for the sum of detectable concentrations of individual pesticides, and 0.03 micrograms per litre for the pesticides aldrin, dieldrin, heptachlor and heptachlor epoxide[27] (persistent organochlorine pesticides now banned in the UK and globally under the Stockholm Convention, but still detected many years after their use). Only 18 'exceedances' of the legal limit for pesticides were detected in 2004. This is a decrease from 31 in 2003 and 72 in 2002. However, data gathered in the PAN *UK* survey (see Table 4) indicates that drinking water contains low levels of pesticides. Looking in detail at both the public and private water supply, the presence of pesticides is confirmed in both raw (untreated), and drinking water at levels which are below the legal limit, but above the limit of detection. As the EC legal limit was intended to achieve zero pesticides in drinking water on a precautionary basis[28], our evidence indicates that, to maintain this objective, the legal limit needs to be lowered to the levels at which pesticides can now be detected. Harmful effects are suspected to risk, in particular, the foetus. The health effects of a lifetime's exposure to a mixture of chemicals at low levels is unknown.

Of the twelve water companies who received and responded to our survey, seven companies completed the section in our questionnaire which asked them to report the levels of pesticides found below the legal limit, but above the limit of detection. In Appendix 6b we present the highest occurrences: atrazine, isoproturon, mecoprop, propyzamide and simazine are all being detected in a high number of tests. According to Southern Water, atrazine occurred in 62.6 percent of drinking water samples, and simazine in 43.2 percent of drinking water samples. Numerous other pesticides are routinely detected in a lower percentage of tests. The water companies varied in their facility to extract data in the format of our information request.

According to the Drinking Water Inspectorate, aldin, dieldrin, heptachlor and heptachlor epoxide are 'generally are not found in water sources[29]'. These are obsolete pesticides, banned since the 1980s on health grounds, which have not had approved uses in the UK for many years. However, the PAN *UK* survey showed that these pesticides are present in water.

Of the respondents to PAN *UK*'s survey who disclosed details about pesticides tested for, and detected at levels below the legal limit, but above the limit of detection, the following water companies reported in their questionnaire return the presence of these pesticides in raw or treated water: Bristol, Dee Valley, Essex

& Suffolk, Mid Kent, Northumbrian, Southern, South Staffordshire, Sutton & East Surrey, Wessex. Aldrin, dieldrin and heptachlor are classified carcinogens and endocrine disruptors according to international authorities[30]. Some studies have found banned pesticides, including these, occurring in rainwater[31] indicating that they persist in the environment in the cycle of precipitation and evaporation.

Private Water Supplies

There are about 42,000 private water supplies in England and 8,000 in Wales, ranging from those supplying a single property to much larger supplies. Although there are some in urban areas, they are mostly in more remote rural parts of the country. The source of the supply may be a well, a borehole, a spring, a lake or a stream[32]. Local authorities have the responsibility for testing for pesticides in private water supplies, see Table.

Only four local authorities disclosed how many pesticides they test for in private water supplies:

North Herefordshire District Council	29
Perth & Kinross Council	25
Penwith District Council	19
Sevenoaks District Council	28

The data on which companies and local authorities base judgements about usage of pesticides on horticultural and most arable crops in their catchment areas is the limited sample survey carried out by the Central Science Laboratory's Pesticide Usage Survey Group[33], which can be analysed to produce locally-relevant data. Some specialist companies produce data on pesticides used on arable crops on a regional basis.

Conclusions

Data gathered in PAN *UK* surveys indicate that pesticides are present in water at low levels. Looking in detail at both the public and private water supply, the presence of pesticides is confirmed in both raw (untreated), and drinking water, at levels which are below the legal limit, but above the limit of detection. As the EC threshold was intended to achieve zero pesticides in drinking water on a

Table 5: Pesticides in Private Drinking Water Supplies		
PAN *UK* Questionnaires sent to:	Approx 468 local authorities in England, Wales, Scotland and Northern Ireland	
Number of responses received (from local authorities with private water supplies):	In 2004 (2002 data) 37	In 2005 (2004 data) 20, 17 of which are from new respondents
Number range of private water supplies for which local authority responsible for testing	1 to 1780 (average 206)	0 to 1153 (average 177)
Number range of private water supplies tested by local authority respondents	1992 to 2002: 0 to 76	In 2003-2004: 0 to 16
Number range of pesticides reported by local authority respondents as tested for	0 to 89	19 to 29
Number of local authority respondents reporting that 100 percent of the pesticides they tested for were detected in all tests above limit of detection	8	Not answered.
Number of local authority respondents reporting exceedances of legal limit (0.1 micrograms per litre)	3	0

precautionary basis[34], our evidence indicates the need to lower the limit. Proper protocols for testing are needed; and mandatory pesticide usage reporting should be implemented to provide a reliable source of information for water companies and local authorities.

4. Discussion

Since the last report there have been a number of important developments at both a national and international level. We review below the most significant for UK pesticide regulation, in terms of human health, in 2005. The evidence of exposure we have presented in this report should be considered in this wider context.

The new report by the Royal Commission on Environmental Pollution, 'Crop spraying and the health of residents and bystanders', has created a potential for change in pesticides policy. It is the most authoritative UK report on the health risks of pesticide exposure for at least fifteen years, and PAN *UK* welcomes its thorough analysis. However the recently published government response has been disappointing[35]. The new All Party Parliamentary Group on Pesticides and Organophosphates was formed in 2005.

A finding in the RCEP report is that illnesses reported by people exposed to pesticides do not match the symptoms that might be anticipated from toxicological tests on laboratory animals. This supports the information collected routinely by PAN *UK* when people report their exposures and symptoms. The latest discoveries in toxicology, especially in relation to the insidious effects of endocrine (hormone) disrupting chemicals, indicate that there must be changes to regulatory toxicology. The true costs of 'endless tests' must be disclosed and the issue openly debated.

The public should be informed by government about the current coverage of tests and its costs, and gaps where there is no knowledge. There should be public participation in the approvals of pesticides and in decision-making about testing, and its costs. The process should be open to scrutiny not only by government regulators, the scientific community and the agrochemical industry, but also by civil society.

The Need for Regulatory Reform

The RCEP has confirmed concerns expressed for many years by PAN *UK*. It identified that 'the PSD combines both delivery of the pesticide approval process and policy advice to Ministers on pesticides', and that 'there is a danger of a conflict of interest, which may be greater where funding is derived from outside government[36].' 'An executive agency of the government, the PSD is funded by government for its policy work, however the full costs of evaluating applications for pesticides approval are recovered from the industry through fees and levies. In the year 2003/04 the PSD received £4.363 million from the levy for regulatory work which includes monitoring and compliance and £2.791 million in industry fees for evaluating applications... (In the same year the PSD received) £5.379 million from Defra for policy-related activities.'[37] PAN *UK* welcomes the long-overdue recognition that these arrangements have profound implications for the governance of pesticides.

We support the RCEP recommendation that government bodies should not hold responsibility for policy and for its execution on the same issue... (and that) these issues should be separated between a government department and an arm's length executive agency or non-departmental public body[38].

The Need for Biomonitoring and Health Outcome Surveillance

The RCEP says 'We were surprised to find that no efforts have been made to establish a database of baseline information for agricultural pesticides that are commonly used in the UK. The principle behind comparing an individual's level with the population norm... is an entirely standard method of proceeding in many areas of clinical diagnosis. Baseline information is being collated in other countries, notably in North America and Germany, and could be used as a framework for information that could be collected in the UK... data on levels of exposure in the population would allow comparison with biomarker levels in an individual subject and provide an understanding of whether the level is unusual and in a range that might lead to an adverse effect. This information could be compared to symptoms of ill health and analysed for trends.[39]'

PAN *UK* has advocated a biomonitoring programme along the lines of the US National Health and Nutrition Examination Survey (NHANES)[40] to regulators for many years. We followed up the RCEP recommendation by advocating at a recent PAN Europe annual conference that a Europe-wide biomonitoring programme and long-term health outcome surveillance programme should be started.

The RCEP identifies that the weakness in most epidemiological studies on pesticides is the 'considerable difficulty of quantifying exposure and identifying the particular pesticide or co-formulant mixture concerned[41].' Formerly one of the most senior government toxicologists in the UK, Dr. Tim Marrs considers that 'Epidemiology is usually of little value in pesticide evaluation', and that the impact on pesticide regulation [for single pesticides] of epidemiological studies has been small, with the possible exception of 2,4,5-T in the 1970s[42]. Dr. Marrs subsequently commented that if a serious effect was occurring post-marketing, something might be expected to be observed in a large number of epidemiology studies. The Advisory Committee on Pesticides describes the system for the review of epidemiological studies carried out by its Medical and Toxicology Panel as one of the key measures in place to check for possible adverse effects once a pesticide has been approved[43].

At the Root of Failing Risk Assessment: Inadequate Toxicity Testing

Questions are now being asked in the most respected scientific journals about the effectiveness and relevance of toxicological tests on laboratory animals which form the basis of current regulatory safety assessment of chemicals including pesticides. For example, the science journal *Nature*, reporting a new initiative by the European Commission to develop alternatives to animal testing acknowledges the poor quality of most animal tests, that they are 'wasteful and poorly predictive', and are 'stuck in a time warp.'[44] According to the report, Thomas Hartung, head of the European Centre for the Validation of Alternative Methods (ECVAM) in Italy said, 'toxicity tests that have been used for decades are simply bad science.' Nature's senior European correspondent Alison Abbott remarks that the experiments have 'never undergone the rigours of validation that in vitro alternatives now face. Most animal tests over- or under-estimate toxicity, or simply don't mirror toxicity in humans very well.'

The current system has not always detected the evidence scientists have discovered in ecological phenomena of the harmful effects of chemicals. The recent Prague Declaration on Endocrine Disruption[45] (Appendix 8), which over 200 scientists have signed, states that 'the existing safety assessment framework for chemicals is ill-equipped to deal with endocrine disrupters. Testing does not account for the effects of simultaneous exposure to many chemicals and may lead to serious underestimations of risk. A fundamental element of chemical safety assessment is the assumption of a threshold dose below which there are no effects. This may not be tenable when dealing with endocrine disrupters, because certain hormonally active chemicals act in concert with natural hormones already present in exposed organisms. Thus even small amounts of chemicals may add to the overall effects, irrespective of thresholds that might exist for these chemicals in the absence of natural hormones. Additionally, due to limited sensitivity of established test methods, it is likely that effects are overlooked.'

The RCEP concurs with the signatories of The Prague Declaration in advocating the development of new assays and screening methods which should 'take advantage of modern technologies such as genomics, proteomics, bioinformatics and metabonomics': 'the element of uncertainty inherent in using animal models might be reduced by the development of the new integrated and molecular based technologies, such as the use of toxicogenomic methods or human cell culture models and the development of animal models of multisymptom/multisystem disease. Toxicogenomics is a tool in development, but it has the potential to inform and improve risk assessment in the future. Under strictly controlled experimental conditions, human cell culture models can be used to re-create human cellular function in an in vitro environment. This could reduce the need for animal models and animals for experimentation and deserves further exploration for those involved in pesticide regulation[46]. There is, however, considerable scientific disagreement on the how accurately different forms of testing can predict outcomes[47].

The relevance of animal data to human disease is questionable when considering the wide inter- and intra-species variability of just one toxicological endpoint, the lethal dose. Data from the same species but different studies are significantly variable and the ten-fold uncertainty factor between species is barely adequate in the case of some pesticides, for example, chlorpyrifos. In the case of diuron, there is a ten-fold variability within studies. For lambda-cyhalothrin, there can be a three or four-fold variability up to a 30-fold variability, depending on which study is selected.

There is currently no regulatory mechanism to ensure that human dose-effect data held by the National Poisons Information Service (and equivalent organisations internationally) from human poisonings, are fed into the pesticides approval process, although PAN *UK* welcomes new initiatives of ECVAM, which recognises the critical importance of human data: '...Validation—the proof that a test accurately predicts a specific effect in humans—is the biggest challenge for alternative methods. One of the 40 or so tests now going through validation is the new cytotoxicity test to help replace the animal lethal dose (LD50) test. It was the first validation study to involve both US and European groups from the start. It is also the first to use data from the records at national poison centres. The predictions of the in vitro test provided a better match than the rat LD50 test when compared with the toxicity information on 42 chemicals listed has having poisoned people.'[48]

There are serious ethical concerns about laboratory animal testing, and the inefficiency and wastefulness of the current system is referred to in the Prague Declaration: 'It is regrettable that commercial pressures and property rights often stand in the way of making publicly available the data gathered by industrial companies for the purposes of hazard identification. We propose that the relevant data from animal testing should be made publicly available whenever possible. This would avoid costly duplication of experiments, and take account of ethical issues ensuring that the best use can be made of animal data for the development of alternative tests.'

Gaps in Testing

New concern has been expressed recently about shortcomings in the testing regime in relation to the supposedly non-active ingredients in pesticide formulations. According to French scientist Gilles-Eric Seralini[49]:

> Scientific problems do exist in the registration of pesticides today, when chronic toxicity tests are conducted with the active ingredient alone—which is generally the case. First of all, chemists from companies may work hard for several years to find the formulation which best amplified the effects of the active ingredient. This formulation will allow penetration and stability and/or bioaccumulation of the active ingredients within plant, fungi or insect cells, for instance, to reach the best toxicity. If there are any side effects in other animal or human cells, these will be also amplified by adjuvants, and thus not measured in chronic toxicity tests with the active ingredient alone. The active compound absorption by skin is generally calculated in the presence of formulated adjuvants, but this is clearly a short-term study and not sufficient to detect, for example, endocrine disruption or carcinogenesis, possibly promoted in vivo by the described synergy. This should even necessitate further care in the use of formulated products such as glyphosate-based herbicides on tolerant, edible plants.

Secret Practice: Human Pesticide Testing

Over the last year there has been considerable high level debate initiated in the US about this issue. There is pressure on the agrochemical industry globally to conduct these studies 'to reduce uncertainty' because there is increasing demand for more sophisticated safety assessment. In the US, since 1996, the Food Quality Protection Act requires the Environmental Protection Agency to add an additional uncertainty factor of between 2 and 10 to account for the special susceptibility of infants and children to toxic substances, unless there are data to the contrary.

PAN North America has now launched a campaign with Earthjustice and the Natural Resources Defense Council challenging the practice[50].

Human pesticide testing is unregulated in the UK. For historical reasons it is exempt from new stringent regulation imposed on experimental trials of medicinal pharmaceuticals. However, there is considerable public concern. The internationally accepted legal instrument controlling human pesticide testing is the Declaration of Helsinki 1964, which introduced an ethical framework to be applied to all biomedical research on human beings, to prevent the reoccurrence of the evils of Nazi experimentation. Current[51] human pesticide testing is conducted in complete secrecy within the private sector, and the lack of scientific and ethical scrutiny to which trials of pharmaceutical drugs are subject is a matter of concern.

5. Conclusions and Recommendations

The need for a more precautionary approach to the approvals and use of pesticides is now imperative, and a national strategy for a government-led, coordinated reduction in the use of these toxic substances is urgently needed to protect human health and the environment.

People continue routinely to be exposed to pesticides in food, water and the environment, reflecting the lack of progress in reducing dependence on these toxic compounds in agriculture. Despite the trend towards lower-dose pesticides that are used in smaller quantities than previously, the total amount of pesticides used in the UK in agriculture in 2004 rose to over 31,000 tonnes. The new national strategy must set clear targets for a coordinated reduction of use.

The costs to human health cannot currently be accurately assessed. A new surveillance system is needed which is quick and easy to access, and provides prompt and efficient investigation and biochemical analysis. Medical outcomes should be monitored.

The existing government schemes should be analysed and amalgamated so numbers of people whose health has been affected by exposure to pesticides can be accurately assessed. It is currently not possible to estimate how many exposures occur because each scheme does not indicate whether or not cases have also been reported to the others. The health effects of chronic exposures to low doses of pesticides, to which we are all subject, are uncertain.

PAN *UK* urges the government to make the following changes to achieve a reduction in people's pesticide exposure; improvements in the governance of pesticides; the strengthening of post-approvals human health surveillance for pesticide-related disease; and the provision of public information. We want to see a reduction in use of 50 percent by 2015, and major reductions in exposure via food, water and the environment.

The Government Should

Reduce People's Pesticide Exposure

1. Include in the national pesticides strategy an action plan for the protection of public health. High levels of pesticide use inevitably result in continuing exposure of people through residues in food, water, and in the environment. The national strategy should include clear targets for reduction. A comprehensive national pesticide usage reporting scheme should be incorporated and there should be disclosure of sales and usage records.
2. Press for a lowering of the European Commission legal limit for pesticides in water in line with the current advances in scientific limits of detection and application of precautionary standards.
3. Adopt and implement fully the recommendations in the Prague Declaration (Appendix 8, Shortcomings of the current regulatory framework, and Proposed measures and actions to be taken) in its own policy and practice and through its civil servants, employees, consultants, contractors and others as relevant.

Improve the Governance of Pesticides

4. Carry out an independent review, possibly through the current Hampton review process, of the ACP and PSD, involving the Department of Health and the Health Protection Agency, and open up their procedures to public scrutiny. Complete restructuring of the regulatory authorities should be considered. .Public health must be prioritised above chemical pest control, and the Department of Health/HPA should be given a more powerful remit in relation to pesticide-related disease.

5. Prevent conflicts of interest in the decision-making process on pesticide testing and approval by separating the functions of pesticide policy making from the pesticide approvals procedures, which receive a proportion of its funds from the agrochemical industry.

6. Introduce and publish, as part of the national pesticides strategy, standardised national protocols for the selection of laboratory analyses of pesticides in food, water and air, for the use of water companies, local authorities, the Drinking Water Inspectorate, the Pesticide Residues Committee, and academic and research institutions engaged in this work.

7. Review the approvals of pesticides which are repeatedly detected in water supplies, or detected after a specific number of occurrences, and if residues in water cannot be avoided their approval should be revoked.

8. Provide opportunities for public consultation in decision-making about the extent of testing and its resourcing. There should be a window for public consultation in the approvals process for every pesticide, as there exists for genetically modified organisms.

9. Oblige agrochemical and veterinary medicines companies trading in UK to disclose full details and data from tests conducted on human subjects of any substance approved in the UK anywhere in the world since 2000, so that their ethical and scientific implications can be scrutinised by policy-makers, parliament, and the public, and regulatory control introduced.

Strengthen Post-approvals Human Health Surveillance

10. Act on the Royal Commission's recommendations and to introduce an effective surveillance scheme for pesticide-related disease, to be run by the Health Protection Agency (HPA) (and the equivalent organisations in the devolved administrations), the cost of which (estimated at £5-10 million per year) should be covered by a levy on pesticides sales. The HPA and related organisations in the devolved administrations should collect population data on pesticides and other chemicals and their biomarkers suspected to cause chronic diseases. A national pesticide and chemical exposure database should be initiated.

11. Press within the European Parliament and Commission for a Europe-wide biological monitoring programme of pesticides and chemicals in humans, in conjunction with the Health Protection Agency's remit above.
12. Implement a programme of primary prevention of diseases linked with pesticides—including cancer and also neurological disease, reproductive disease, immunological disease, respiratory disease, and skin disease—through a national programme of exposure reduction to toxic substances and the adoption of an adequately funded and supported toxics use reduction strategy, including a programme to phase out pesticides which are known to be relatively hazardous.

Provide Public Information

13. Give residents, walkers, and all other countryside users the right to know what pesticides they are being exposed to by introducing mandatory notification of use both in advance and with signs on site.
14. Initiate a scientific and public debate on the potential benefits of reduced chemical use.

(Alison Craig, Pesticide Action Network UK, She can be reached at alisoncraig@pan-uk.org).

APPENDIX 1

Pesticide Usage Trends in the UK 1992-2004

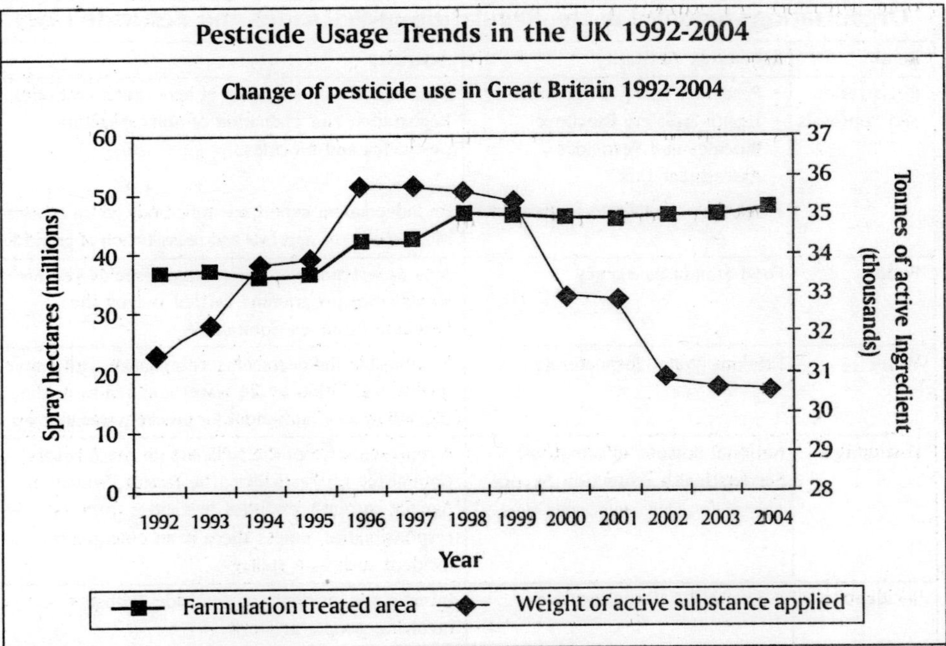

Total pesticide usage in Great Britain, 2004 (by weight of active ingredient): 30.6 thousand tonnes; (by area treated) 47.7 million hectares.

'Treated area' is the gross area treated with a pesticide, including all repeat applications.

Source: Central Science Laboratory Pesticide Usage Survey Group (PUSG) http://pusstats.csl.gov.uk/ All the recent PUSG survey reports are available as pdf documents at: http://www.csl.gov.uk/science/organ/pvm/puskm/pusg.cfm.

APPENDIX 2

Organisations Involved in the Regulation of Pesticides and Pesticide Laws

Remit	Regulatory Authority	Activities
Registration and approvals	• Pesticides Safety Directorate • Health & Safety Executive Biocides and Pesticides Assessment Unit • Advisory Committee on Pesticides	Registration and evaluation of agricultural pesticides. Registration and evaluation of non-agricultural pesticides and biocides. An independent expert scientific body which advises Ministers on the approval and re-evaluation of pesticides.
Food	Food Standards Agency	Acts as watchdog in overseeing pesticide residue surveillance programme carried out by the Pesticide Residues Committee.
Water	Drinking Water Inspectorate	Responsible for overseeing compliance with water quality legislation by 24 water companies in the UK; and by local authorities for private water supplies.
Poisonings	National Poisons Information Service/Health Protection Agency	A representative of the NPIS sits on the Advisory Committee on Pesticides. The Health Protection Agency currently excludes pesticides from its responsibilities, unless there is an emergency incident such as a spillage.
'Incidents'	• Health & Safety Executive • Local Authority Environmental Health Departments	Investigates incidents of pesticide exposure involving people at work Investigates incidents of pesticide exposure not involving people at work
Air	National Air Emissions	Inventory Monitors toxins in air, but only includes 3 pesticides, all obsolete: lindane (gamma-HCH), PCP (pentachlorophenol), and HCB (hexachlorobenzene)

See also:

HSE leaflet: Reporting incidents of exposure to pesticides and veterinary medicines INDG141(rev1) 2/99 C1000.

A guide to pesticide regulation in the UK and the role of the Advisory Committee on Pesticides, ACP 19 (311/2005), Department for the Environment, Food and Rural Affairs, and the Health & Safety Executive, undated, *www.pesticides.gov.uk* ACP homepage.

Pesticide Laws

The use, supply, storage and advertisement of pesticides is regulated by a number of pieces of legislation including, for Great Britain, the Control of Pesticides Regulations (COPR) and Plant Protection Products Regulations (PPPR). PSD is responsible for agricultural pesticides; most non-agricultural pesticides are the responsibility of Health and Safety Executive (HSE). PPPR is the newer legislation and implements a European Directive (91/414/EEC) which regulates 'Plant Protection Products'; these include agricultural pesticides and growth regulators.

The use of pesticides is also regulated by COSHH (the Control of Substances Hazardous to Health).

For a full guide to the legislation see the UK Pesticide Law section on the PSD website *www.pesticides.gov.uk*

APPENDIX 3

The Regulatory Testing and Assessment of Pesticides

According to the official guidance* pesticide legislation aims to ensure that

- Pesticides are only approved for use if they are effective;
- No-one develops any serious illness through the use of pesticides;
- No-one is harmed or made ill by the presence of pesticide residues in food or drink;
- When pesticides are used according to the conditions of their approval, any adverse effects on wildlife or the environment are sufficiently small to be deemed acceptable.

At the approval stage, regulators make a careful scientific assessment of the pesticide, based mainly on laboratory animal experiments. In 2004, 28,252 procedures on 28,243 animals relating to 'agriculture' were carried out, involving animals**. This excludes tests done on animals for non-agricultural pesticides, biocides, and household products.

Human toxicity data is used when occupational health and epidemiological studies are available, but before a substance is registered there has been very little, if any, human exposure. Regulators periodically review pesticides in use and carry out the surveillance of pesticide-related ill-health via a number of schemes, aiming to detect adverse effects which were not anticipated at the approval stage.

For a pesticide to be registered, the agrochemical company has to submit to the regulators a 'dossier' comprising an extensive range of data, broadly falling into seven areas:

1. Physico-chemical properties
2. Potential toxicity in humans
3. Dietary intake
4. Exposure to operators, other workers and 'bystanders'
5. Environmental fate and behaviour
6. Ecotoxicology
7. Efficacy and risk to following crops.

To assess the potential toxicity in humans of the pesticide, the scientists aim to establish a 'No Adverse Effect Level' (NOAEL) for any ill-effects that might occur. A NOAEL is the highest dose administered to laboratory animals in an investigation that does not cause them 'observable' ill-effects. Nonobservable effects, such as subtle changes in neurological function, pain, dizziness, numbness and psychological symptoms, are not included in this value.

The data that are required to assess potential human toxicity cover:

- How the active ingredient is metabolised and excreted in mammals.
- The 'acute' toxicity of a single high dose of the active ingredient and of the product by oral, dermal and inhalation exposure, usually in rats.

Contd...

Contd...

- The 'sub-acute' and 'chronic' toxicity of the active ingredient when administered to animals over periods of several weeks or longer (in two species, typically rats for up to two years and dogs for up to a year).
- The potential of the active ingredient to cause cancer when it is administered over a lifetime (in two species, usually in rats for a minimum of two years and in mice for 18 months).
- The genotoxicity of the active ingredient, i.e. its potential to damage the genetic material in cells, potentially causing cancer. Genotoxicity is (exceptionally) treated as a hazard trigger. Genotoxic pesticides are now not registered.
- The developmental toxicity of the active ingredient, i.e. whether it can cause birth defects when administered to female animals during pregnancy.
- The toxicity of the active ingredient when it is administered to at least two successive generations of animals over the course of their lifetime. This provides further information on the chronic toxicity of the pesticide and aims to detect its potential to impair fertility and the ability to rear young.
- The potential of the active ingredient and product to irritate the skin or eyes.
- The potential of the active ingredient and product to cause skin allergies (sensitisation).
- Further tests may be required if there is a need to understand effects better, for example, on particular organ systems such as the nervous, immune or endocrine systems.

On the basis of these data, regulators decide whether the product needs to be labelled as a hazard (for example, irritant, harmful, toxic). 'Acceptable' levels of exposure are also defined, expressed as the Acceptable Daily Intake (ADI), Acute Reference Dose (ARfD), Acceptable Operator Exposure Level (AOEL) and others.

Acceptable Daily Intake (ADI)

This is the amount of a chemical which can be consumed every day for a lifetime in the practical certainty, on the basis of all known facts, that no harm will result. It is expressed in milligrams of the chemical per kilogram bodyweight of the consumer. The starting point for the derivation of the ADI is usually the lowest 'no adverse effect level' (NOAEL) that has been observed in animal studies of toxicity. This is then divided by an uncertainty factor (most often 100) to allow for the possibility that animals may be less sensitive than humans and also to account for possible variation in sensitivity between individuals. The studies from which NOAELs and ADIs are derived are supposed to take into account any impurities in the pesticide active ingredient as manufactured, and also any toxic breakdown products (metabolites) of the pesticide, but can only do so if and when these have been identified.

Acute Reference Dose (ARfD)

The definition of the ARfD is similar to that of the ADI, but it relates to the amount of a chemical that can be taken at one meal or on one day. It is normally derived by applying an appropriate uncertainty factor to the lowest NOAEL in studies that assess acute toxicity or developmental toxicity.

Contd...

Contd...

Acceptable Operator Exposure Level (AOEL)

This is intended to define a level of daily exposure that would not cause adverse effects in operators who work with a pesticide regularly over a period of days, weeks or months. Depending on the pattern of use, a short-term AOEL (i.e. for exposures over several weeks or on a seasonal basis), or long-term AOEL (i.e., for repeated exposures over the course of a year) are defined. AOELs are normally derived from a short-term laboratory animal toxicity study or a multi-generation study.

Maximum Residue Level (MRL)***

Pesticide residues in food are controlled through UK regulations which lay down maximum residue levels. MRLs are not safety limits for residues in food. They are designed to check that pesticides are being used correctly, according to good agricultural practice (GAP). In order to avoid serious inconsistencies in MRLs between countries, the Codex Alimentarius Commission (a United Nations body) has established the Codex Committee on Pesticide Residues, which bases its work on the scientific approvals made by the Food and Agriculture (FAO)/World Health Organisation (WHO) Joint Meeting on Pesticide Residues****. Where there is no approved use of a pesticide nor an 'import tolerance' a residue level is set at the limit of determination (LOD) (an affective 'zero' reflecting the lowest level at which reliable quantitative analysis can be performed).

Sources

* A guide to the regulation of pesticides in the UK and the role of the Advisory Committee on Pesticides, ACP 14 (299/2003), Department for the Environment, Food and Rural Affairs and Health & Safety Executive, June 2003, *www.pesticides.gov.uk*

** Home Office, Statistics of Scientific Procedures on Living Animals in Great Britain, 2004, Stationery Office.

*** Annual report of the Pesticide Residues Committee 2002.

**** Control of pesticides and IPM, Implementation of Farmer Participatory Integrated Pest Management and Better Chemical Management in ACP States, Directorate-General for Development, Commission of the European Communities, Pesticides Trust, June 1998.

APPENDIX 4A

'The Human Health Incidents Survey': Pesticide Exposures and Poisonings Reported to the Pesticides Safety Directorate (PSD) by Companies									
Type	**User Type**	**Year**						**Total Number of Cases**	**Percentage of Cases**
		2002		**2003**		**2004**			
		No.	**%**	**No.**	**%**	**No.**	**%**		
Involving one or more children	Amateur Use	12	9%	20	13%	24	14%	56	12%
	Professional Use	7	5%	3	2%	2	1%	12	3%
	Unknown	1	1%					1	0%
Involving one or more children (Total)		**20**	**15%**	**23**	**15%**	**26**	**15%**	**69**	**15%**
Other	Amateur Use	58	42%	79	52%	75	42%	212	45%
	Professional Use	57	42%	49	32%	73	41%	179	38%
	Unknown	2	1%	1	1%	3	2%	6	1%
Other Total		**117**	**85%**	**129**	**85%**	**151**	**85%**	**397**	**85%**
Grand Total		**137**	**100%**	**152**	**100%**	**177**	**100%**	**466**	**100%**

APPENDIX 4B

Pesticide Poisonings Reported to the National Health Service (NHS)

From the National Poisons Information Service Combined Annual Report 2004-05

Types of products which are subject of enquiries to NPIS	Percentage of Enquiries
Agrochemicals inc home use	1%
Animals	<1%
Chemicals, industrial	12%
Cosmetics	2%
Household	8%
Others	7%
Pharmaceuticals	67%
Plants/fungi	3%

Reports on Poisoning with Selected Agents (Pesticides) (2004/05)

Pesticide	Number of telephone enquiries out of a total of 113,125	Place of exposure	Route of exposure	Poisoning Severity Score at time of enquiry, when recorded
Carbamate insecticides	89	91% in the home; 1% in agricultural workplace; 6.7% deliberate 2 chronic exposures;	69% ingestion; 19% inhalation; 6% skin contact; 6% eye contact.	61.8% no symptoms; 33.2% minor symptoms. No moderate or severe exposures and no deaths reported.
Glyphosate	120	81.7% in the home; <1% in agricultural workplace; 13.3% deliberate exposures.	33% ingestion; 24% inhalation; 31% skin contact; 7% eye contact; 3% injection; 1% multiple; 1% other.	25.7% no symptoms, 60.7% minor symptoms; 12.5% moderate. None severe. None reported to have long-term sequelae. One death acutely.
Organo-phosphorus insecticides	137	78.8% in home; 3% in agricultural workplace; 7.3% deliberate exposure.	55% ingestion; 21% inhalation; 12% skin contact; 9% eye contact; 3% multiple; 0% other.	54.2% no symptoms; 44.4% minor symptoms; no moderate and 1% severe. No deaths or long-term sequelae reported.
Paraquat	125	16% in agricultural workplace; 61.6% in the home; 20% deliberate exposures.	50% ingestion; 12% inhalation; 29% skin contact; 4% eye contact; 1% injection; 0% multiple; 4% other.	33.8% no symptoms; 47.8% minor symptoms; 11.9% moderate; 1.5% severe; 2 deaths reported.

Contd...

Contd...				
Pyrethroids	268	90% in the home; 4.1% in agricultural workplace; 4.5% deliberate exposures; 5 chronic exposures.	54% ingestion; 26% inhalation; 8% skin contact; 7% eye contact; 0% injection; 1% multiple; 4% other.	38.4% no symptoms; 59.4% minor symptoms; 0.7% moderate; 1.4% severe, both related to eye contact. No deaths reported.
Total enquiries for carbamate insecticides, glyphosate, organophosphorus insecticides, paraquat and pyrethroids.	739			
Note: No data on TOXBASE (online) sessions are disclosed.				

APPENDIX 5A

Food Residues (Europe)

Results of the eighteen national monitoring programmes for pesticide residues on fresh (including frozen) fruit, vegetables and cereals, sum of surveillance and enforcement samples.

Country	Number of Samples Analysed		% of Samples without Detectable Residues		% of Samples with Residues below or at MRL		% of Samples with Confirmed Residues above the MRL		% of Samples with Multiple Residues	
	2002	2003	2002	2003	2002	2003	2002	2003	2002	2003
Belgium	1028	1250	55	55	40	41	2.5	4.2	15.2	12.1
Denmark	1977	1530	60	54	38	43	2.4	2.9	17.7	22.8
Germany	7035	10586	46	43	45	49	5.1	8.4	31.1	32.1
Greece	1661	1659	56	77	42	21	1.9	2.2	7.8	5.4
Spain	4049	3670	62	66	35	30	3.5	4.5	8.6	7.9
France	3721	3372	47	50	44	43	6.2	7.0	29.9	23.7
Ireland	617	1022	52	59	44	37	4.2	3.5	18.2	12.6
Italy	8095	7172	70	69	28	29	1.1	1.7	14.0	13.3
Luxembourg	118	107	60	50	36	48	1.7	1.9	11.9	11.2
Netherlands	3042	2549	46	42	38	44	8.2	14.4	31.1	33.7
Austria	1637	1404	46	69	38	27	8.2	4.0	29.2	13.5
Portugal	722	363	74	61	23	30	2.8	9.4	9.6	12.9
Finland	1985	1725	49	55	46	38	4.1	6.7	27.7	22.4
Sweden	2073	2131	58	50	37	43	4.0	6.9	17.7	22.4
United Kingdom	2087	2452	56	66	43	33	1.6	1.0	20.7	15.9
Norway	2280	2164	66	63	30	34	3.4	2.3	15.5	17.7
Iceland	278	313	53	61	45	38	2.5	1.3	23.7	8.9
Liechtenstein	47	47	81	70	17	28	0.0	2.1	0.0	4.3

MRL = Maximum Residue Level (legal trade limit, not safety limit)

Source: Monitoring of pesticide residues in products of plant origin in the European Union, Norway, Iceland and Lichtenstein, 2002 and 2003 (latest data, November 2005).

Exceedances of Acute Reference Dose Levels

Exposure assessment for acute risk from the pesticides investigated in the 2003 coordinated programme for the products with the highest residues found in a composite sample in the European Union. The calculation was performed with the UK Exposure Model for an adult (70.1 kg) and a toddler (15.5 kg) and only those pesticides which have acute toxicity and where an acute Reference Dose has been set.

Food item sampled, 2002	Pesticide	Exceedances: intake as percentage of Acute Reference Dose	Food item sampled, 2003	Pesticide	Exceedances: intake as percentage of Acute Reference Dose
Potatoes	Aldicarb	151% (toddler)	Cucumber	Oxydemeton-methyl	400% (toddler)
Beans	Methamidophos	477% (toddler)	Grapes	Chlorpyrifos	147% (toddler)
Beans	Methiocarb	381% (toddler)	Grapes	Dimethoate	112% (toddler)
Carrots	Diazinon	103% (toddler)	Grapes	Lambda-cyhalothrin	334% (toddler)
Peach	Acephate	160% (toddler)	Grapes	Methomyl	257% (toddler)
Peaches	Parathion	161% (toddler)	Grapes	Methomyl	1035% (toddler)
Carrots	Aldicarb	134% (toddler)	Sweet peppers	Endosulfan	217% (toddler)
Oranges	Methidathion	125% (toddler)	Sweet peppers	Methamidophos	164% (toddler)
Oranges	Triazophos	393% (toddler)	Sweet peppers	Methiocarb	142% (toddler)
Spinach	Methomyl	116% (adult), 351% (adult), 456% (toddler)	Sweet peppers	Triazophos	187% (adult), 507% (toddler)
Spinach	Oxydemeton-methyl	102% (adult), 310% (adult), 404% (toddler)			

APPENDIX 5B

Food Residues (UK Testing Programme)					
Results Reported in the Pesticide Residues Committee Monitoring Reports, 2004.					
Food	**Pesticide**	**WHO Hazard**	**OP WHO Class**	**Cancer**	**Endocrine Disrupting Chemicals**
Apples	Azinphos-methyl	Acutely toxic WHO Ib	OP		
	Captan			Suspected carcinogen USEPA B2; EU 3; IARC 3	
	Carbaryl	Acutely toxic WHO II		Suspected carcinogen USEPA 2; EU 3; IARC 3	Suspected endocrine disrupting chemical EU1
	Carbendazim			Suspected carcinogen USEPA C	Suspected endocrine disrupting chemical EU2
	Chlorpyrifos	Acutely toxic WHO II	OP		
	Diphenylamine				
	Dithianon				
	Dithio-carbamates				
	Dodine				
	Folpet				
	Iprodione			Suspected carcinogen USEPA L2; EU 3	Suspected endocrine disrupting chemical EU2
	Malathion		OP	Suspected carcinogen USEPA 3; IARC 3	Suspected endocrine disrupting chemical EU2
	Metalaxyl				
	Myclobutanil				
	Paclobutrazol				
	Phosalone	Acutely toxic WHO II	OP		
	Phosmet	Acutely toxic WHO II	OP	Suspected carcinogen USEPA 3	
	Pirimicarb	Acutely toxic WHO II			
	Propargite				

Contd...

Contd...					
	Thiabendazole			Suspected carcinogen USEPA 2, 4	
	Tolyfluanid			Suspected carcinogen USEPA 2	
Asparagus	Cypermethrin	Acutely toxic WHO II		Suspected carcinogen USEPA C	Suspected endocrine disrupting chemical EU2
Beans, speciality	Captan			Suspected carcinogen USEPA B2; EU 3; IARC 3	
	Carbendazim			Suspected carcinogen USEPA C	Suspected endocrine disrupting chemical EU2
	Chlorpyrifos	Acutely toxic WHO II	OP		
	Cypermethrin	Acutely toxic WHO II		Suspected carcinogen USEPA C	Suspected endocrine disrupting chemical EU2
	Deltamethrin	Acutely toxic WHO II		Suspected carcinogen IARC 3	Suspected endocrine disrupting chemical EU1
	Dicofol			Suspected carcinogen USEPA C; IARC 3	Suspected endocrine disrupting chemical EU2; OSPAR
	Dimethoate	Acutely toxic WHO II	OP	Suspected carcinogen USEPA C	Suspected endocrine disrupting chemical UK EA; EU2
	Dithiocar-bamates				
	Metha-midophos	Acutely toxic WHO Ib	OP		
	Methomyl	Acutely toxic WHO Ib			Suspected endocrine disrupting chemical EU2
	Omethoate	Acutely toxic WHO Ib	OP		
	Profenofos	Acutely toxic WHO II	OP		
	Propargite				
	Tetradifon				
	Triazophos	Acutely toxic WHO Ib			
Beer	Chlormequat				
					Contd...

Contd...					
Bread, ordinary	Chlormequat				
	Glyphosate			Suspected carcinogen USEPA B2; EU 3	
	Malathion		OP	Suspected carcinogen USEPA 3; IARC 3	Suspected endocrine disrupting chemical EU2
	Pirimiphos-methyl		OP		
	Chlormequat				
	Glyphosate			Suspected carcinogen USEPA B2; EU 3	
	Malathion		OP	Suspected carcinogen USEPA 3; IARC 3	Suspected endocrine disrupting chemical EU2
Cabbage, head	Iprodione			Suspected carcinogen USEPA L2; EU 3	Suspected endocrine disrupting chemical EU2
	Metalaxyl				
	Triazamate	Acutely toxic WHO II			
Carrots	Iprodione			Suspected carcinogen USEPA L2; EU 3	Suspected endocrine disrupting chemical EU2
Chillies	Carbendazim			Suspected carcinogen USEPA C	Suspected endocrine disrupting chemical EU2
	Carbofuran	Acutely toxic WHO Ib			Suspected endocrine disrupting chemical EU2
	Chlorpyrifos	Acutely toxic WHO II	OP		
	Cypermethrin	Acutely toxic WHO II		Suspected carcinogen USEPA C	Suspected endocrine disrupting chemical EU2
	Dicofol			Suspected carcinogen USEPA C; IARC 3	Suspected endocrine disrupting chemical EU2; OSPAR
	Dimethoate	Acutely toxic WHO II	OP	Suspected carcinogen USEPA C	Suspected endocrine disrupting chemical UK EA; EU2
	Dimethoate	Acutely toxic WHO II	OP	Suspected carcinogen USEPA C	Suspected endocrine disrupting chemical UK EA; EU2
	Fenvalerate	Acutely toxic WHO II		Suspected carcinogen IARC 3	Suspected endocrine disrupting chemical EU2
					Contd...

Contd...

	Iprodione			Suspected carcinogen USEPA L2; EU 3	Suspected endocrine disrupting chemical EU2
	Metalaxyl				
	Metha-midophos	Acutely toxic WHO Ib	OP		
	Myclobutanil				
	Omethoate	Acutely toxic WHO Ib	OP		
	Profenofos	Acutely toxic WHO II	OP		
	Trifloxy-strobin				
Citrus, soft	2,4-D	Acutely toxic WHO II			
	2-phenyl-phenol				
	Bromo-propylate				
	Carbendazim			Suspected carcinogen USEPA C	Suspected endocrine disrupting chemical EU2
	Carbofuran	Acutely toxic WHO Ib			Suspected endocrine disrupting chemical EU2
	Chlorpyrifos	Acutely toxic WHO II	OP		
	Dicofol			Suspected carcinogen USEPA C; IARC 3	Suspected endocrine disrupting chemical EU2; OSPAR
	Dimethoate	Acutely toxic WHO II	OP	Suspected carcinogen USEPA C	Suspected endocrine disrupting chemical UK EA; EU2
	Diphenylamine				
	Ethion	Acutely toxic WHO II OP			
	Fenthion	Acutely toxic WHO II OP			
	Imazalil	Acutely toxic WHO II		Suspected carcinogen USEPA L1	
	Iprodione			Suspected carcinogen USEPA L2; EU 3	Suspected endocrine disrupting chemical EU2

Contd...

Contd...					
	Malathion		OP	Suspected carcinogen USEPA 3; IARC 3	Suspected endocrine disrupting chemical EU2
	Methidathion	Acutely toxic WHO Ib	OP	Suspected carcinogen USEPA C	
	Omethoate	Acutely toxic WHO Ib	OP		
	Prochloraz				
	Propargite				
	Pyriproxifen				
	Thiabendazole			Suspected carcinogen USEPA 2, 4	
Fish, farmed	Chlordane	Acutely toxic WHO II		Suspected carcinogen USEPA B2; EU 3; IARC 2B	Ger EA; EU1; OSPAR
	DDT	Acutely toxic WHO II		Suspected carcinogen USEPA B2; EU 3; IARC 2B	Suspected endocrine disrupting chemical UK EA; Ger EA; EU1; OSPAR
	Dieldrin			Suspected carcinogen USEPA B2; EU 3; IARC 3	Suspected endocrine disrupting chemical UK EA; EU2; OSPAR
	Hexachloro-benzene	Acutely toxic WHO Ia		Suspected carcinogen USEPA B2; EU 2; IARC 2B	
Grapes	Azoxystrobin				
	Bromopropylate				
	Captan			Suspected carcinogen USEPA B2; EU 3; IARC 3	
	Carbaryl	Acutely toxic WHO II		Suspected carcinogen USEPA 2; EU 3; IARC 3	Suspected endocrine disrupting chemical EU1
	Carbendazim			Suspected carcinogen USEPA C	Suspected endocrine disrupting chemical EU2
	Chlorpyrifos	Acutely toxic WHO II	OP		
	Chlorpyrifos-methyl		OP		
	Cypermethrin	Acutely toxic WHO II		Suspected carcinogen USEPA C	Suspected endocrine disrupting chemical EU2
	Dimethoate	Acutely toxic WHO II	OP	Suspected carcinogen USEPA C	Suspected endocrine disrupting chemical UK EA; EU2
					Contd...

Contd...					
	Dithio-carbamates				
	Fenitrothion	Acutely toxic WHO II	OP	Suspected carcinogen EU1	
	Imazalil	Acutely toxic WHO II		Suspected carcinogen USEPA L1	
	Iprodione			Suspected carcinogen USEPA L2; EU 3	Suspected endocrine disrupting chemical EU2
	Lambda-cyhalothrin	Acutely toxic WHO II			
	Metalaxyl				
	Methomyl	Acutely toxic WHO Ib			Suspected endocrine disrupting chemical EU2
	Mono-crotophos	Acutely toxic WHO Ib	OP		
	Omethoate	Acutely toxic WHO Ib	OP		
	Pirimiphos-methyl		OP		
	Procymidone				Suspected endocrine disrupting chemical EU2
	Pyrimethanil			Suspected carcinogen USEPA C	
	Tebuconazole			Suspected carcinogen USEPA 3	
	Trifloxystrobin				
	Vinclozolin			Suspected carcinogen USEPA C; EU 3	Suspected endocrine disrupting chemical Ger EA; EU1; OSPAR
Infant food (meat/egg/fish/cheese)	Chlorpropham			Suspected carcinogen IARC 3	
Kiwi fruit	Carbaryl	Acutely toxic WHO II		Suspected carcinogen USEPA 2; EU 3; IARC 3	Suspected endocrine disrupting chemical EU1
	Carbendazim			Suspected carcinogen USEPA C	Suspected endocrine disrupting chemical EU2
					Contd...

Contd...					
	Iprodione			Suspected carcinogen USEPA L2; EU 3	Suspected endocrine disrupting chemical EU2
	Procymidone				Suspected endocrine disrupting chemical EU2
	Vinclozolin			Suspected carcinogen USEPA C; EU 3	Suspected endocrine disrupting chemical Ger EA; EU1; OSPAR
Leeks	Azoxystrobin				
	Chlorothalonil				
	Phosalone	Acutely toxic WHO II	OP		
Lettuce	Acephate		OP	Suspected carcinogen USEPA C	Suspected endocrine disrupting chemical EU2
	Azoxystrobin				
	Cypermethrin	Acutely toxic WHO II		Suspected carcinogen USEPA C	Suspected endocrine disrupting chemical EU2
	Cyprodinil				
	Dimethoate	Acutely toxic WHO II	OP	Suspected carcinogen USEPA C	Suspected endocrine disrupting chemical UK EA; EU2
	Dithiocar-bamates				
	Endosulfan	Acutely toxic WHO II			UK EA; EU2; OSPAR
	Fenitrothion	Acutely toxic WHO II	OP	Suspected carcinogen EU1	
	Fludioxanil				
	Folpet				
	Imidacloprid	Acutely toxic WHO II			
	Inorganic bromide				
	Iprodione			Suspected carcinogen USEPA L2; EU 3	Suspected endocrine disrupting chemical EU2
	Lambda-cyhalothrin	Acutely toxic WHO II			
	Metha-midophos	Acutely toxic WHO Ib	OP		
					Contd...

Contd...

	Procymidone				Suspected endocrine disrupting chemical EU2
	Propamocarb				
	Propyzamide			Suspected carcinogen EU 3	
	Pyrimethanil			Suspected carcinogen USEPA C	
	Toclofos-methyl				
	Vinclozolin			Suspected carcinogen USEPA C; EU 3	Suspected endocrine disrupting chemical Ger EA; EU1; OSPAR
Nuts	Inorganic bromide				
Oats/rye	Chlormequat				
	Chlorpyrifos-methyl		OP		
	Glyphosate			Suspected carcinogen USEPA B2; EU 3	
	Mepiquat				
	Pirimiphos-methyl		OP		
Okra	Chlorpyrifos	Acutely toxic WHO II OP			
	Cyfluthrin	Acutely toxic WHO II			
	Cypermethrin	Acutely toxic WHO II		Suspected carcinogen USEPA C	Suspected endocrine disrupting chemical EU2
	Dimethoate	Acutely toxic WHO II	OP	Suspected carcinogen USEPA C	Suspected endocrine disrupting chemical UK EA; EU2
	Endosulfan	Acutely toxic WHO II			UK EA; EU2; OSPAR
	Ethion	Acutely toxic WHO II	OP		
	Mono-crotophos	Acutely toxic WHO Ib	OP		
	Omethoate	Acutely toxic WHO Ib	OP		
	Propargite				

Contd...

Contd...					
Parsnips	Aldicarb	Acutely toxic WHO Ia		Suspected carcinogen IARC 3	Suspected endocrine disrupting chemical EU2
	Chlorfen-vinphos		OP		Suspected endocrine disrupting chemical EU2
	Trifluralin			Suspected carcinogen USEPA C; IARC 3	Suspected endocrine disrupting chemical UK EA; EU2?
Pears	Azinphos-methyl	Acutely toxic WHO Ib	OP		
	Captan			Suspected carcinogen USEPA B2; EU 3; IARC 3	
	Carbaryl	Acutely toxic WHO II		Suspected carcinogen USEPA 2; EU 3; IARC 3	Suspected endocrine disrupting chemical EU1
	Carbendazim			Suspected carcinogen USEPA C	Suspected endocrine disrupting chemical EU2
	Chlormequat				
	Chlorpyrifos	Acutely toxic WHO II	OP		
	Diphenylamine				
	Dithio-carbamates				
	Dodine				
	Folpet				
	Imazalil	Acutely toxic WHO II		Suspected carcinogen USEPA L1	
	Iprodione			Suspected carcinogen USEPA L2; EU 3	Suspected endocrine disrupting chemical EU2
	Malathion		OP	Suspected carcinogen USEPA 3; IARC 3	Suspected endocrine disrupting chemical EU2
	Metalaxyl				
	Phosmet	Acutely toxic WHO II	OP	Suspected carcinogen USEPA 3	
	Pirimicarb	Acutely toxic WHO II			
	Procymidone				Suspected endocrine disrupting chemical EU2
	Thiabendazole			Suspected carcinogen USEPA 2, 4	
					Contd...

Contd...					
	Tolyfluanid			Suspected carcinogen USEPA 2	
Peas	Chlorothalonil				
	Deltamethrin	Acutely toxic WHO II		Suspected carcinogen IARC 3	Suspected endocrine disrupting chemical EU1
	Dimethoate	Acutely toxic WHO II OP	OP	Suspected carcinogen USEPA C	Suspected endocrine disrupting chemical UK EA; EU2
	Dithio-carbamates				
	Metha-midophos	Acutely toxic WHO Ib	OP		
	Omethoate	Acutely toxic WHO Ib	OP		
	Tebuconazole			Suspected carcinogen USEPA 3	
	Triadimenol			Suspected carcinogen USEPA C	Suspected endocrine disrupting chemical EU2
Plantain	Carbendazim			Suspected carcinogen USEPA C	Suspected endocrine disrupting chemical EU2
	Imazalil	Acutely toxic WHO II		Suspected carcinogen USEPA L1	
	Thiabendazole			Suspected carcinogen USEPA 2, 4	
Potatoes	Aldicarb	Acutely toxic WHO Ia		Suspected carcinogen IARC 3	Suspected endocrine disrupting chemical EU2
	Chlorpropham			Suspected carcinogen IARC 3	
	Imazalil	Acutely toxic WHO II		Suspected carcinogen USEPA L1	
	Maleic hydrazide			Suspected carcinogen IARC 3	
	Oxadixyl			Suspected carcinogen USEPA C	
	Tecnazene				
Pulses	Acephate		OP	Suspected carcinogen USEPA C	Suspected endocrine disrupting chemical EU2
					Contd...

Contd...

	Carbaryl	Acutely toxic WHO II		Suspected carcinogen USEPA 2; EU 3; IARC 3	Suspected endocrine disrupting chemical EU1
	Chlorpyrifos	Acutely toxic WHO II	OP		
	Hydrogen phosphide				
	Inorganic bromide				
	Metha-midophos	Acutely toxic WHO Ib	OP		
Salad, pre-packed	Acephate		OP	Suspected carcinogen USEPA C	Suspected endocrine disrupting chemical EU2
	Azoxystrobin				
	Bifenthrin	Acutely toxic WHO II		USEPA C	Suspected endocrine disrupting chemical EU1
	Cypermethrin	Acutely toxic WHO II		Suspected carcinogen USEPA C	Suspected endocrine disrupting chemical EU2
	Cyprodinil				
	Dicloran				
	Dimethoate	Acutely toxic WHO II	OP	Suspected carcinogen USEPA C	Suspected endocrine disrupting chemical UK EA; EU2
	Endosulfan	Acutely toxic WHO II			UK EA; EU2; OSPAR
	Fenitrothion	Acutely toxic WHO II	OP	Suspected carcinogen EU1	
	Fenvalerate	Acutely toxic WHO II		Suspected carcinogen IARC 3	Suspected endocrine disrupting chemical EU2
	Folpet				
	Imidacloprid	Acutely toxic WHO II			
	Iprodione			Suspected carcinogen USEPA L2; EU 3	Suspected endocrine disrupting chemical EU2
	Lambda-cyhalothrin	Acutely toxic WHO II			
	Metha-midophos	Acutely toxic WHO Ib	OP		

Contd...

Contd...					
	Oxadixyl			Suspected carcinogen USEPA C	
	Procymidone				Suspected endocrine disrupting chemical EU2
	Pyrimethanil			Suspected carcinogen USEPA C	
	Tebuconazole			Suspected carcinogen USEPA 3	
	Vinclozolin			Suspected carcinogen USEPA C; EU 3	Suspected endocrine disrupting chemical Ger EA; EU1; OSPAR
Straw-berries	Azoxystrobin				
	Bifenthrin	Acutely toxic WHO II		USEPA C	Suspected endocrine disrupting chemical EU1
	Bupirimate				
	Captan			Suspected carcinogen USEPA B2; EU 3; IARC 3	
	Carbendazim			Suspected carcinogen USEPA C	Suspected endocrine disrupting chemical EU2
	Chlorothalonil				
	Chlorpyrifos	Acutely toxic WHO II	OP		
	Clofentezine			Suspected carcinogen USEPA C	
	Cyprodinil				
	Dicofol			Suspected carcinogen USEPA C; IARC 3	Suspected endocrine disrupting chemical EU2; OSPAR
	Dithio-carbamates				
	Endosulfan	Acutely toxic WHO II			UK EA; EU2; OSPAR
	Fenpropimorph				
	Iprodione			Suspected carcinogen USEPA L2; EU 3	Suspected endocrine disrupting chemical EU2
	Kresoxim-methyl			Suspected carcinogen USEPA L1; EU 3	
					Contd...

Contd...					
	Mepanipyrim				
	Myclobutanil				
	Pirimicarb	Acutely toxic WHO II			
	Procymidone				Suspected endocrine disrupting chemical EU2
	Pyrimethanil			Suspected carcinogen USEPA C	
	Tetradifon				
	Tolyfluanid			Suspected carcinogen USEPA 2	
	Triadimenol			Suspected carcinogen USEPA C	Suspected endocrine disrupting chemical EU2
Sweet pepper	Azoxystrobin				
	Chlorothalonil				
	Cypermethrin	Acutely toxic WHO II		Suspected carcinogen USEPA C	Suspected endocrine disrupting chemical EU2
	Dithio-carbamates				
	Endosulfan	Acutely toxic WHO II			UK EA; EU2; OSPAR
	Fludioxanil				
	Imidacloprid	Acutely toxic WHO II			
	Iprodione			Suspected carcinogen USEPA L2; EU 3	Suspected endocrine disrupting chemical EU2
	Malathion		OP	Suspected carcinogen USEPA 3; IARC 3	Suspected endocrine disrupting chemical EU2
	Metalaxyl				
	Methomyl	Acutely toxic WHO Ib			Suspected endocrine disrupting chemical EU2
	Oxamyl	Acutely toxic WHO Ib			
	Pirimiphos-methyl		OP		
					Contd...

Contd...

	Procymidone				Suspected endocrine disrupting chemical EU2
	Tebufenpyrad			Suspected carcinogen USEPA 3	
Tomatoes	Azoxystrobin				
	Bifenthrin	Acutely toxic WHO II		USEPA C	Suspected endocrine disrupting chemical EU1
	Bupirimate				
	Buprofezin				
	Carbendazim			Suspected carcinogen USEPA C	Suspected endocrine disrupting chemical EU2
	Chlormequat				
	Chlorothalonil				
	Cypermethrin	Acutely toxic WHO II		Suspected carcinogen USEPA C	Suspected endocrine disrupting chemical EU2
	Cyprodinil				
	Deltamethrin	Acutely toxic WHO II		Suspected carcinogen IARC 3	Suspected endocrine disrupting chemical EU1
	Dicofol			Suspected carcinogen USEPA C; IARC 3	Suspected endocrine disrupting chemical EU2; OSPAR
	Difenoconazole			Suspected carcinogen USEPA C	
	Endosulfan	Acutely toxic WHO II			UK EA; EU2; OSPAR
	Fenbatatin				
	oxide				
	Fenhaxamid				
	Fludioxanil				
	Iprodione			Suspected carcinogen USEPA L2; EU 3	Suspected endocrine disrupting chemical EU2
	Kresoxim-methyl			Suspected carcinogen USEPA L1; EU 3	
	Mepanipyrim				
	Oxadixyl			Suspected carcinogen USEPA C	
	Procymidone				Suspected endocrine disrupting chemical EU2

Contd...

Contd...					
	Propargite				
	Pyrimethanil			Suspected carcinogen USEPA C	
	Tebuconazole			Suspected carcinogen USEPA 3	
	Tebufenpyrad			Suspected carcinogen USEPA 3	
	Tolyfluanid			Suspected carcinogen USEPA 2	
	Triadimenol			Suspected carcinogen USEPA C	Suspected endocrine disrupting chemical EU2
No residues were found in tests of corn on the cob, coffee, marmelade, milk, orange juice, sweetcorn (canned), beef, tuna (canned), turkey and cheese (mature and mild) in 2004.					

Hazard to Health Issues

World Health Organisation Classifications					
Class		**LD_{50} for the rat (mg/kg body weight)**			
		Solids (Oral)	**Liquids**	**Solids (Dermal)**	**Liquids**
Ia	Extremely hazardous	5 or less	20 or less	10 or less	40 or less
Ib	Highly hazardous	5-50	20-200	10-100	40-400
II	Moderately hazardous	50-500	200-2000	100-1000	400-4000
III	Slightly hazardous	Over 500	Over 2000	Over 1000	Over 4000
U	Unlikely to present acute hazard in normal use: 'WHO Table 5'				
O	Active ingredients believed to be obsolete or discontinued for use as pesticides.				

The terms 'solid' and 'liquids' refer to the physical state of the active ingredient. The LD_{50} value is a statistical estimate of the number of mg of toxicant per kg of bodyweight required to kill 50% of a large population of test animals.

Endocrine Disrupting Chemicals

UK EA – on the UK Environment Agency's list of target EDCs

Strategy for Endocrine disrupting chemicals, *http://www.environmentagency.gov.uk/commondata/105385/139909*

DEFRA – identified as associated with endocrine disruption by the UK Department for Environment, Food and Rural Affairs, website: Hormone Disrupting Substances in the Environment *http://www.defra.gov.uk/environment/hormone/index.htm*

Contd...

Contd...

Ger.EA – potential and confirmed EDCs by the German Federal Environment Agency column, Pesticides suspected of endocrine-disrupting effects by Germany's Federal Environment Agency, ENDS Report 290, March 1999.

EU – considered as high concern EDC by the European Union, Commission moots priority list of endocrine chemicals, BKH/TNO report, June 2000.

OSPAR – identified as a potential EDC under Oslo and Paris Commission, Endocrine disrupting pesticide: Gwynne Lyons. Pesticides News 46, December 1999.

Definitions of Cancer Categories

US Environmental Protection Agency

The US EPA has changed its classification systems in recent years. Some categories have similar definitions:

Weight-of-evidence categories developed during the 1980s

Group B = Probable Human Carcinogen: B1 indicates limited human evidence; *B2* indicates sufficient evidence in animals and inadequate or no evidence in humans.

Group C = Possible Human Carcinogen:

Weight-of-evidence categories developed during the 1990s

Known/Likely available tumour effects and other key data are adequate to demonstrate convincingly a carcinogenic potential for humans.

L1 = Likely at high doses but Not Likely at low doses

L2 = Likely to be carcinogenic to humans, available tumour effects and other key data are adequate to demonstrate carcinogenic potential for humans.

S = Cannot be Determined-Suggestive evidence from human or animal data is suggestive of carcinogenicity, but is not sufficient to conclude as to human carcinogenic potential.

Source: Office of Pesticide Programs List of Chemicals Evaluated for Carcinogenic Potential, US EPA, [see details at www.epa.gov/pesticides/carlist/ although list not available on website], August 2000.

European Union

There is no single EU list available denoting carcinogenic pesticides. EC Directive 67/548 and subsequent amendments provide the classification of dangerous substances, including pesticides. The cancer classifications are:

Category 2 (denoted as R45 on the pesticide label) = May Cause Cancer

Category 3 (denoted as R40 on label) = *Possible Risk of Irreversible Effects (Cancer*, as cited in table)

Sources: EC Directive 67/548 EEC and subsequent amendments; Chemicals (Hazard Information and Packaging for Supply) [CHIP2] Regulations 1994, Health and Safety Executive, UK.

International Agency for Research on Cancer

Group 1 = Carcinogenic to Humans

Group 2A = Probably Carcinogenic to Humans (limited evidence of carcinogenicity in humans and sufficient evidence in experimental animals).

Group 2B = Possibly Carcinogenic to Humans (limited evidence of carcinogenicity in humans and less than sufficient evidence in experimental animals).

Source: http://193.51.164.11/monoeval/grlist.html [Note: lists cited include many non-pesticides].

APPENDIX 6A

Pesticides in the Public Drinking Water Supply

From PAN UK Survey, 2005 (2004 Data).

Water company	Total number of pesticide determinands* tested for in 2004	Failures for pesticides in 2004, and level(s) found (EC Drinking Water Directive limit 0.1 micrograms per litre)**	Pesticide removal costs: 1) capital cost to date 2) operational cost for 2004	In your Company's view, should overall pesticide usage be reduced?
Anglian	19 separate pesticide determinands with 18147 tests in total	Trietazine, 7: 0.12, 0.11, 0.11, 0.13, 0.12, 0.12, 0.11	Not available in this format.	There are advantages and disadvatages of the use of pesticides and a balance needs to be sort [sic].
Bournemouth & West Hampshire	N/A	N/A	N/A	N/A
Bristol	32, 9,364 [tests]	0	Not specifically available: ozone and GAC [granular activated carbon] treatment is installed for taste and odour control in addition to pesticide removal.	It is the Company's policy to minimise the use of pesticides and herbicides and only to use non-persistent types.
Cambridge	N/A	N/A	N/A	N/A
Cholderton & District	N/A	N/A	N/A	N/A
Dee Valley	25, 1326 on final waters	MCPA, 1: 0.2	Not given.	'Not known.'
Dwr Cymru	N/A	N/A	N/A	N/A
Essex & Suffolk	39 pesticide determinands tested for; 6980 tests carried out on raw water sources; 9243 tests on final waters for regulatory compliance purposes	0	Not available.	We do not feel this is a matter for water companies to comment upon.
Folkestone & Dover	N/A	N/A	N/A	N/A
Hartlepool	5 separate pesticide determinands with 100 tests in total	0	Nil	There are clearly advantages and disadvatages of the use of pesticides and a balance needs to be sort [sic].
Mid Kent	48	0	Not given.	Not answered.

Contd...

Contd...

Northumbrian	49 pesticide determinands tested for; 11,115 tests carried out on raw water sources; 27,354 tests on final waters for regulatory compliance purposes	MCPA, 5: 0.12, 0.17, 0.19, 0.20, 0.24	Not given.	We are not in a position to comment on this matter.
Portsmouth	25,2482 [tests]	2,4-D, 1: 0.133; metaldehyde, 1: 0.23; Simazine, 1: 0.658; total pesticides, 1: 0.677	Not available.	Not answered.
Severn Trent	N/A	N/A	N/A	N/A
South East	N/A	N/A	N/A	N/A
South Staffordshire	45,9839 tests	0	Not given.	Yes.
South West	N/A	N/A	N/A	N/A
Southern	Raw: 23,134 tests for 54 pesticides; treated: 22,443 tests for 39 pesticides.	2,4-D, 1: 0.15	Capital cost to date: £46 million; operational cost for 2004: ~£225,000.	There are statutory obligations for the control of certain weeds which will require the correct use of pesticides. Proper use and management of pesticide applications should reduce any risk to the water supply and environment.
Sutton & East Surrey	12 individual pesticides and total pesticides. Total number of determinations = 1339 (raw and treated).	0	Not given.	Not answered.
Tendring Hundred	N/A	N/A	N/A	N/A
Thames Water Utilities	N/A	N/A	N/A	N/A
Three Valleys	N/A	N/A	N/A	N/A
United Unitilies	N/A	N/A	N/A	N/A
Wessex	26	0	Not available.	Pesticides should be more carefully selected. Atrazine, isoproturon, [and] mecoprop should be removed from use when better, less mobile alternatives available.
Yorkshire	N/A	N/A	N/A	N/A

* Determinands = specific parameter for analysis

** Source = Drinking Water Inspectorate Annual Report 2004

N/A = not available as no questionnaire returned.

APPENDIX 6B

PANUK Survey: Pesticides in Public Drinking Water Supplies – Worst Offenders

Percentage of samples (raw and treated, i.e. drinking water) in which specified pesticides were detected above the limit of detection.

Water Company	Total Pesticides	Atrazine	Isoproturon	MCPP Mecoprop	Propyzamide	Simazine
Bristol	64%	79%	15.7%	19.6%	N/A	63.7%
Essex & Suffolk	N/A	31%	33%	25%	N/A	54%
Mid Kent	N/A	40%	0	9%	0	31%
Northumbrian	N/A	27%	11%	2%	100%	24%
South Staffordshire	79.5%	79.7%	7%	55.4%	0	59.8%
Southern	N/A	62.4% raw; 62.6% treated	5.5% raw; 1.2% treated	10.5% raw; 3.8% treated	4.6% raw	47.7% raw; 43.2% treated
Sutton & East Surrey	89%	85%	6%	14%	N/A	34%

Apart from Southern, water companies did not specify if pesticide detected in raw or treated water.

APPENDIX 8

Prague Declaration on Endocrine Disruption, May 2005

Summary

- There is serious concern about the high prevalence of reproductive disorders in European boys and young men and about the rise in cancers of reproductive organs, such as breast and testis. Lifestyle, diet and environmental contamination play a role in the observed regional differences of these disorders and their changes with time.
- Hormone action is important in the origin or progression of the aforementioned disorders. Therefore it is plausible that exposure to endocrine disrupters may be involved, but there are inherent difficulties in establishing such causal links in humans.
- There is a serious gap of knowledge regarding the effects of endocrine disruptive compounds on other serious human diseases such as obesity, neuronal disorders, stress etc.
- Causality is well established for detrimental effects in wildlife as a direct consequence of exposure to endocrine disrupters. In some instances the severity of effects is likely to lead to population level impacts. Wildlife provides early warnings of effects produced by endocrine disrupters which may as yet be unobserved in humans.
- Wildlife represents a protection target in its own right. The severity of endocrine disrupting effects observed in the laboratory indicates that these substances may pose a threat for wildlife biodiversity as already shown for organotin compounds and marine snails.
- Europeans are exposed to low levels of a large number of endocrine disrupters which can act in concert. Many of these chemicals, drugs or natural products are found in human tissues and in breast milk. Humans are exposed to these chemicals from very early on in their lives when the developing organism can be particularly sensitive.
- The existing safety assessment framework for chemicals is ill-equipped to deal with endocrine disrupters. Testing does not account for the effects of simultaneous exposure to many chemicals and may lead to serious underestimations of risk.
- The current safety testing guidelines are based on reproductive effects, and thus do not take into account the deleterious effects of endocrine disruptors in other tissues. New test systems need to be developed to solve this shortcoming.
- In view of the magnitude of the potential risks associated with endocrine disrupters, we strongly believe that scientific uncertainty should not delay precautionary action on reducing the exposures to and the risks from endocrine disrupters.
- The challenges posed by endocrine disrupters require a long-term commitment to monitoring and research which is dedicated to characterising human and wildlife exposure and their mechanisms of action and interaction. This will help ensure better protection of the health of European citizens and the environment.

Contd...

Contd...

Introduction

International experts and scientists representing many different disciplines came together in Prague on 10-12 May 2005 for a workshop on chemicals that interfere with hormone systems, so-called endocrine disrupters. The workshop was convened to discuss recent European research on the health risks associated with these chemicals. Much of this work emanated from large research projects funded by the European Union, and joined together in the cluster for research on endocrine disrupters, CREDO. The results presented at the Prague workshop have reinforced concerns over the long-term consequences of exposure to endocrine disrupters to humans and wildlife.

Endocrine disrupters are a very diverse group of chemicals, including some pesticides, bulk produced chemicals, flame retardants, agents used as plasticisers, cosmetic ingredients, pharmaceuticals, natural products such as plant-derived estrogens and many more. These substances may alter the function of hormonal systems and cause adverse effects by mimicking the effects of natural hormones, blocking their normal action, or by interfering with the synthesis and/or excretion of hormones. The following position statement was agreed by the undersigned scientists. This document is intended to update European citizens, policy makers and regulators on research progress, to highlight shortcomings and flaws in current regulation and to make constructive suggestions that might lead to better protection of human and wildlife health in Europe and beyond.

Research Updates: Human Health Concerns

1. We are concerned about the high prevalence of male reproductive disorders in some European countries. There have been rises in genital malformations in baby boys, and recent data indicate that in parts of Europe, sperm quality is approaching crisis levels that may impair fertility.
2. The incidence of cancers, such as breast, testis and prostate cancer, continues to increase in many European countries, although there are notable differences between countries. The descendants of people who have migrated between countries adopt the cancer incidence of their new home country. This shows that these cancers are linked to factors in the environment, including the diet.
3. Genital malformations, testis cancer, and some cases of reduced sperm quality arise early in life, even during development in the womb. These conditions have common causes during reproductive organ development in the fetus, which is controlled by hormones. The concern is that endocrine disrupters may interfere with these processes to disturb male genital development during pregnancy. Similarly, hormonal dysregulation may lead to the formation of breast cancer in women and abnormal pubertal development in girls.
4. The immune system of young children can be affected by exposure to polychlorinated biphenyls (PCBs) and dioxins during development in the womb. As a result, the likelihood of contracting infectious diseases is increased. PCBs and dioxins are hormonally active

Contd...

Contd...

pollutants found in the diet. These substances degrade very slowly, accumulate in fatty tissue and are able to reach the developing fetus. After birth, they are passed to babies via mother's milk. We are concerned that these contaminants, at levels found in food, induce unwanted health effects in young children. Steroid and thyroid hormones are involved in brain development and brain ageing and many other effects. Environmental contaminants which affect these systems may increase the risk of brain dysfunction.

5. Although at this point there is no clear link between exposure to thyroid hormone disruptors, cancer, mental retardation has reduced fertility and neurodegenerescence in humans these issues need urgent evaluation, as such problems can be expected from our basic knowledge of the thyroid hormone physiology.
6. Little or no information is currently available regarding the effects of endocrine disrupters on disease condition outside the reproductive system such as metabolic syndrome, neuronal development, childhood cancers, cognitive development, immune problems, psychological disorders learning and memory development, and other. In many cases there are casual links between endocrine disrupters and these diseases and more scientific information is required.
7. Use of novel research technologies in understanding the mechanisms of endocrine disruptor action at the molecular level is required. By understanding the molecular mechanisms that are affected by endocrine disruptors it will be easier to extrapolate the information between different exposed tissues.

Research Updates: Linking Cause and Effect

8. Undoubtedly, European citizens have experienced rises in reproductive disorders and hormone-dependent cancers. What is unclear, however, is whether these diseases are linked to exposure to endocrine disrupters. Establishing a link is difficult as human diseases are the result of many interacting influences, of which chemicals are but one determinant. Only when a chemical exerts a very strong impact, has it been possible to uncover its role in disease processes, as is the case with steroidal estrogens and breast cancer. It is much more difficult, to prove small, albeit existing, influences of chemicals on health. Thus, we are convinced that failure to demonstrate direct links between hormone-related disorders and exposure to chemicals should not be taken to indicate an absence of risks.
9. The identification of causative chemicals is complicated by the possibility that disorders may become manifest long after exposure has taken place. By this time, causative agents may have disappeared from tissues, thus obscuring identification of risk.

Research Updates: Wildlife Effects

10. Beyond the fact that wildlife represents a protection target in their own right, they act as sentinels for effects produced by endocrine disrupters which may as yet be unobserved in humans. Seals living in the Baltic Sea and the North Sea have suffered reproductive

Contd...

Contd...

failure and population declines that can be attributed to the impact of PCBs and dioxins. The same chemicals can also affect the immune system of seals, making them more vulnerable to infection with viruses.

11. Across Europe, male fish exposed to sewage treatment discharges show abnormal levels of female egg yolk protein due to the presence of endocrine disrupters such as steroidal estrogens and surfactant breakdown products in the sewage effluent. Reproductive abnormalities in fish, notably the appearance of eggs in the testes of male fish have also been observed. These fish have been shown to have a reduced reproductive capacity and 'males' produce sperm of poorer quality. Negative impacts on entire fish populations may be the consequence, as has been shown in recent laboratory studies on several fish species. Fish exposed to the contraceptive pill ingredient at concentrations found in European rivers showed disturbed sexual development and impaired reproductive capabilities at the adult stage, including reduced or inhibited egg production and egg fertilisation hindered release of semen and lower survival of their offspring.
12. Invertebrates are also vulnerable to the effects of endocrine disrupters. Tributyl tin, an ingredient in antifouling paints applied to the hulls of ships, resulted in the formation of male sex organs in female molluscs, with consequent reductions in population numbers. More recently, it was shown that bisphenol A, an industrial chemical, and UV-filter substances utilised in sun screens cause increased egg production in aquatic snails. The consequences of such abnormalities for the balance and well-being of entire ecosystems are not yet predictable, but the severity of effects observed indicates a potential impact on wildlife biodiversity from endocrine disrupters.

Research Updates: Exposure

13. Considerable progress has been made in identifying new endocrine active chemicals. These include chemicals used as UV filters and antioxidants in cosmetics and chemicals used as preservatives in food. It is clear that European citizens are simultaneously exposed to large numbers of endocrine disrupters. However, we do not know the full range that we are exposed to through our diets, drinking water, air and consumer products. This lack of knowledge severely hampers efforts to explore a link between exposure and resultant effects in humans.
14. Human tissue levels of PCBs and dioxins have stabilised at approximately one third of the pollution peak in the 1970s. This indicates that internal exposure to these substances will continue, with European populations having to live with a pollution burden that will be present for generations to come.
15. We are concerned that Europe is currently experiencing an increase in pollution with highly persistent brominated chemicals that are used as flame retardants in many consumer items, including furnishings and computers. These substances and their breakdown products are found in mother's milk, food items, wildlife and many environmental

Contd...

Contd...

media. Current knowledge regarding the exposures as well as the toxicological profile of these chemicals are insufficient for a proper human and ecological risk assessment.

Research Updates: Safety Testing and Regulation

16. A fundamental element of chemical safety assessment is the assumption of a threshold dose below which there are no effects. This may not be tenable when dealing with endocrine disrupters, because certain hormonally active chemicals act in concert with natural hormones already present in exposed organisms. Thus, even small amounts of chemicals may add to the overall effects, irrespective of thresholds that might exist for these chemicals in the absence of natural hormones. Additionally, due to limited sensitivity of established test methods, it is likely that effects are overlooked.
17. A further complication is that hormonal effects are often masked by other toxic responses. Only when testing is carried out at low doses usually not administered during routine testing do these effects become apparent. Furthermore, a feature of endocrine disrupters is the late occurrence of adverse effects long after exposure has ceased. Existing testing methods are not generally designed to deal with this possibility.
18. These difficulties are exacerbated when the effects of simultaneous exposure to many chemicals (mixture effects) are considered. Recent studies have shown that mixture effects can occur even when each component is present at a dose that individually does not produce effects. These observations further undermine the belief that threshold doses can be applied meaningfully during the safety assessment of chemicals. A dose of a single chemical judged to be safe after testing in isolation may give a false sense of security when exposure includes large numbers of other endocrine active chemicals which may interact with each other.

Shortcomings of the Current Regulatory Framework

19. The array of standardised methods that exists for the safety assessment of chemicals is ill-equipped to identify endocrine disrupters or to anticipate their likely effects on humans and wildlife. Many pollutants now recognised as endocrine disrupters, such as the case of tributyl tin, and certain phthalates (used as plasticisers in consumer goods), were only identified through scientific studies, not by routine safety testing. By this time considerable environmental damage had already been caused. Therefore, there is an urgent need to improve existing, and to develop novel, regulatory test methods.
20. Due to the weaknesses of existing regulations in identifying endocrine disrupters, biological and chemical monitoring programmes become increasingly important for the detection of as yet unidentified effects missed during the current risk assessment of chemicals. Existing monitoring programmes lack the ability to deal appropriately with endocrine disrupters, and chemical and biological monitoring must exist in concert.
21. Environmental exposure is to a mixture of chemicals and not a single agent. However, this is not reflected in test protocols and provisions to take mixture effects into account

Contd...

Contd...

are totally lacking. Recent research indicates that this may lead to a significant underestimation of risks. The issue is beginning to receive attention among regulators, but jointly, regulators and scientists need to cooperate to develop workable approaches to dealing with mixtures.

22. Current test protocols rely on effects on the male and female reproductive tract. Testing protocols need to be developed to assess the effects of endocrine disruptors in other relevant tissues.

Proposed Measures and Actions to be Taken

23. For the foreseeable future, regulation of endocrine disrupters will have to cope with the tension between the biological plausibility of serious, perhaps irreversible damage and delays in generating data suitable for comprehensive risk assessment. In view of the magnitude of the potential risks, we strongly believe that scientific uncertainty should not delay precautionary action for risk reduction.

24. There are various frameworks to guide decision making about the selection of endocrine disrupters for further testing. Prioritisation is usually achieved by using screening assays to select chemicals for extensive testing which delays regulatory action until further data is available. Though screening assays are not adequate as a basis for risk assessment, they should be utilised to trigger precautionary regulatory action on the basis of the rebuttable assumption that positive results may indicate risks. Precautionary action can include labelling, measures to reduce exposure, restrictions in use patterns or even the ban on certain chemicals.

25. The substances already known to have endocrine disrupting properties should be included in the proposed European chemicals regulation REACH, and subject to the authorisation procedure. Initially, the substances should be drawn from existing lists detailed in the EU strategy for endocrine disrupters. By a dynamic process, it is imperative that new substances should enter and exit the list taking account of new information, particularly including academic studies, as it becomes available.

26. Steps should be taken to restrict inherently the use of persistent chemicals, e.g. brominated flame retardants in order to halt their build-up in humans and the environment. We are concerned that inaction will lead to a dangerous repeat of the events that have led to the accumulation of dioxins and PCB's in humans and wildlife.

27. The release of endocrine disrupters from sewage treatment works should be reduced significantly. A large fraction of the pollution stems from steroid hormones excreted by humans, and the control of these cannot be easily regulated. Therefore, improvements to sewage treatment technology for the removal of these and other endocrine disrupters are required. However, where practicable, for man-made substances, priority should be given to the prevention of the release, rather than end of pipe solutions.

Contd...

Contd...

28. It is regrettable that commercial pressures and property rights often stand in the way of making publicly available the data gathered by industrial companies for the purposes of hazard identification. We propose that relevant data from animal testing should be made publicly available whenever possible. This would avoid costly duplication of experiments, and take account of ethical issues ensuring that the best use can be made of animal data for the development of alternative tests.

Research Priorities

29. The challenges posed by endocrine disrupters cannot be solved in the short term, and there is an urgent need for further research to underpin better protection of the health of European citizens and the environment. To aid the planning of the forthcoming 7th Framework Programme of EU research funding, we propose that research activities in this area should be prioritised, as follows:
30. The lack of a complete picture of the full array of endocrine disrupters is hampering progress with risk assessment. Further extensive research during the next five to ten years is needed to fill gaps. Emphasis should be placed on the development of new chemical analytical methods and the development and validation of bioassay-directed techniques. Biobanks with suitable human and wildli'fe reference material, covering European countries with marked differences in relevant disorders and/or chemical exposure should be established.
31. Further understanding of the possible modes of action of endocrine disrupters is required in order to recognise organism functions that might be at risk. Only on the basis of such research will it be possible to develop appropriate biomarkers and biotests of effects for human and wildlife disorders. A considerable strengthening of links to fundamental research into disease processes is necessary. The effects of endocrine disruptors on novel target tissues and a wider array of cellular signalling pathways need to be elucidated, in particular those closely linked to disease conditions.
32. The effects of endocrine disruptors on a wider array of cellular signalling pathways needs to be elucidated, in particular those closely linked to disease conditions. Focus should be placed on signalling pathways involved in major disease conditions such as metabolic syndrome, obesity, and heart disease.
33. The development of new assays and screening methods for the identification of endocrine disrupters relevant to humans and wildlife should be pursued with urgency. This should take advantage of modern technologies such as genomics, proteomics, bioinformatics and metabonomics.
34. More mechanistic information regarding how endocrine disruptors are involved in human disease is required. This information need to take into account the complexity of the effect and exposure scenario with multiple targets, exposure to multiple contaminants and the fact that exposure levels are low and exposure time is long.

Contd...

Contd...

35. Further systematic work on mixture effects will be needed to underpin better risk assessment procedures. Research should be extended to exploring relationships between exposure time and dose, and to investigations of the effects of sequential exposure to several chemicals. Emphasis should be placed on understanding the mechanistic basis of combination effects.
36. The consequences of endocrine disruption in wildlife for the balance and well-being of ecosystems should be pursued with urgency because some case studies have already shown that endocrine disrupters pose a threat for biodiversity. Emphasis should be placed on better linkage of laboratory and field investigations, considering a broad coverage of vertebrate and invertebrate groups.
37. In wildlife research, mechanistic work linking effects seen at the organism level to population-level and ecosystems effects should be encouraged. There is a need to apply the rigorous methodology of human epidemiology to the wildlife arena. Links with ecological systems approaches should be encouraged.
38. Special programmes focusing on the detection of possible effects on the newborn child giving rise to problems in childhood and adulthood should be initiated in order to overcome the challenge of possible long temporary breach between exposure episode and overt adverse outcome.

Source: www.edenresearch.info/declaration.html

Glossary of Acronyms

ACP	Advisory Committee on Pesticides
ADI	Acceptable Daily Intake
AOEL	Acceptable Operator Exposure Level
APPG	All Party Parliamentary Group
ARfD	Acute Reference Dose
CSL	Central Science Laboratory
Defra	Department for Food and Rural Affairs (UK)
DWI	Drinking Water Inspectorate
EC	European Commission
EDC	Endocrine Disrupting Chemical
EU	European Union
FSA	Food Standards Agency
HPA	Health Protection Agency
HSE	Health & Safety Executive
IARC	International Agency for Research on Cancer
LA	Local Authority
Mg/kg/bw/d	Milligrams per kilogram of bodyweight per day
MP	Member of Parliament
MRL	Maximum Residue Level
NAEI	National Air Emissions Inventory
NOAEL	No Observable Adverse Effect Level
NPIS	National Poisons Information Service
OP	Organophosphate
PAN UK	Pesticide Action Network UK
PCB	Polychlorinated Biphenyls
PEX	Action on Pesticide Exposure (PAN UK project)
PIAP	Pesticide Incidents Appraisal Panel (HSE's)
PRC	Pesticide Residues Committee
PSD	Pesticides Safety Directorate
RCEP	Royal Commission on Environmental Pollution
UK	United Kingdom
USEPA	United States Environmental Protection Agency
WHO	World Health Organisation

References

1. Craig A, People's Pesticide Exposures – poisons we are exposed to every day without knowing it, PAN *UK*, London, December 2004, *http://www.pan-uk.org/briefing/PeoplesPesticide Exposures.pdf*.
2. Royal Commission on Environmental Pollution, 'Crop spraying and the health of residents and bystanders', September 2005, *www.rcep.org.uk/cropspraying.htm*
3. Environment, Food and Rural Affairs Select Committee, 'Progress on pesticides' report, 5th April 2005.
4. Royal Commission on Environmental Pollution, 'Crop spraying and the health of residents and bystanders', September 2005, *www.rcep.org.uk/cropspraying.htm,* page 109.
5. Pesticide Incidents Report, HSE Field Operations Directorate Investigations, 1 April 2004 – 31 March 2005, *www.hse.gov.uk/fod/pir0405.pdf*
6. PAN *UK*, People's Pesticide Exposures, December 2004, page 15.
7. 'Companies fail to meet reporting requirements', Pesticides News, March 2005, page 3.
8. Pers comm Caroline Kennedy, PSD, telephone conversation with Alison Craig, PAN *UK* 13 December 2005.
9. Pesticides Safety Directorate website *www.pesticides.gov.uk/approvals.asp?id=1675*
10. Rob Edwards, Environment Editor, The Sunday Herald, 28 August 2005 *www.sunday herald.com/51506*
11. Pesticides Safety Directorate *www.pesticides.gov.uk/applicant_guide.asp?id=1456*
12. Health Protection Agency *www.hpa.org.uk/chemicals/npis.htm*
13. Adams R D, Good A M, Bateman D N, 'Pesticide exposure monitoring using NPIS resources April 2004 – March 2005', NPIS Edinburgh.
14. HSE, Reporting incidents of exposure to pesticides and veterinary medicines, INDG151(rev1) 2/00 C1000.
15. Nocolopoulou-Stamati, P, Hens L, Howard, C V, Van Larebeke, N, Cancer as an environmental disease, Kluwer Academic Publishers, 2004 ISBN 1 4020 2019 8.
16. Commission of the European Communities, Monitoring of pesticide residues in products of plant origin in the European Union, Norway, Iceland and Lichtenstein, 2003, Brussels, 26.10.2005 SEC(2005) 1399.
17. PAN Europe, Monitoring of pesticide residues in the European Union – 2003 results, November 2005.

18. Commission of the European Communities, Monitoring of pesticide residues in products of plant origin in the European Union, Norway, Iceland and Lichtenstein, 2003, Brussels, 26.10.2005 SEC(2005) 1399, page 45.
19. Central Science Laboratory *www.csl.gov.uk/prodserv/cons/pesticide/intell/pusg.cfm*
20. Pesticide Residues Committee website *www.pesticides.gov.uk/prc.asp?id=955#What_pesticides_are_analysed_for_*
21. Pesticide Residues Committee, Annual report 2004, page 9.
22. Pesticide Residues Committee, Annual Report 2004, page 25.
23. Pesticide Residues Committee updated working paper on the proposed surveillance programme for 2004, to meeting of 30th July 2003.
24. Telephone conversation between David Mason, CSL, and Alison Craig, PAN *UK* 24th November 2005.
25. Letter from the PRC to Alison Craig, PAN *UK*, 18th May 2005, in response to a request for information under the Environmental Information Regulations/Freedom of Information Act 15th April 2005.
26. Drinking Water Inspectorate, Annual report 2004.
27. Drinking Water Inspectorate, Annual report 2004, page 217.
28. Beaumont, Peter, Pesticides, policies and people – a guide to the issue, 1993, page 42.
29. Drinking Water Inspectorate, Annual report 2004, page 217.
30. PAN *UK*, The list of lists briefing paper, November 2005.
31. Van Maanen JMS, De Vaan MAJ, Veldstra AWF and Hendrix WPAM Pesticides and nitrate in groundwater and rainwater in the province of Limburg in the Netherlands Environmental Monitoring and Assessment, 2001, Vol 72, No 1, 95-114.
32. Drinking Water Inspectorate, Drinking Water in England, 2004, page 457.
33. Central Science Laboratory *www.csl.gov.uk/prodserv/cons/pesticide/intell/pusg.cfm*
34. Beaumont, Peter, Pesticides, policies and people – a guide to the issue, 1993, page 42.
35. The Royal Commission on Environmental Pollution report on crop spraying and the health of residents and bystanders – Government response, 20th July 2006, *www.defra.gov.uk*
36. The Royal Commission on Environmental Pollution report 'Crop spraying and the health of residents and bystanders', 22nd September 2005, *www.rcep.org.uk/cropspraying.htm,* page 85.
37. Ibid.
38. Ibid.

39. The Royal Commission on Environmental Pollution report 'Crop spraying and the health of residents and bystanders', 22nd September 2005, *www.rcep.org.uk/cropspraying.htm,* page 31.

40. Centers for Disease Control and Prevention, National Center for Health Statistics, National Health and Nutrition Examination Survey, *www.cdc.gov/nchs/about/major/nhanes/growthcharts/ch arts.htm*

41. The Royal Commission on Environmental Pollution report 'Crop spraying and the health of residents and bystanders', 22nd September 2005, *www.rcep.org.uk/cropspraying.htm,* page 20.

42. Dr. Tim Marrs, Presentation to the Agchem Forum conference, 3rd October 2005.

43. Advisory Committee on Pesticides, 'A guide to pesticide regulation in the UK and the role of the Advisory Committee on Pesticides', ACP 19 (311/2005), undated, page 16.

44. Nature, Vol 438, Issue no 7065, 10th November 2005, 'More than a cosmetic change'.

45. Prague Declaration on Endocrine Disruption, May 2005, *www.edenresearch.info/declaration.html*

46. The Royal Commission on Environmental Pollution report 'Crop spraying and the health of residents and bystanders', 22nd September 2005, *www.rcep.org.uk/cropspraying.htm,* page 30.

47. European Centre for the Validation of Alternative Methods, Institute for Health & Consumer Protection, European Commission Joint Research Centre, 21020 Ispra (VA), Italy, Alternative (non-animal) methods for chemicals testing: current status and future prospects, Draft of 15/05/02, May 2002.

48. The International Programme on Chemical Safety Human Data Initiative *http://www.who.int/ipcs/publications/methods/human_da ta/en/*

49. Seralini, Gilles-Eric, Universite de Caen, France, letter in Environmental Health Perspectives Volume 113, Number 10, October 2005, page A658.

50. Website of PAN North America, *www.panna.org*

51. Email response from Syngenta to Rob Edwards, Sunday Herald: 'In some cases data is needed from volunteers to refine the assessment and/or provide additional data to be used in the risk assessment of established products', in email from Rob Edwards to Alison Craig, PAN *UK,* 6th October 2004.

5

Pesticide Use and Pesticide Residues*

D Atkinson, F Burnett, G N Foster, A Litterick, M Mullay and C A Watson

'Pesticide Use and Pesticide Residues' has been excerpted from The Minimisation of Pesticide Residues in Food: A Review of the Published Literature. The published information on pesticide residues indicates that a significant proportion of most major vegetable crops and all major fruit products consumed in the US (& Europe) contain residues of pesticides. Pesticide residues detected in samples analyzed show commonly consumed fruits, vegetables and processed foods, considered as health foods are not free of pesticide residues. A whole host of vegetables and 'healthy' fruits and commonly consumed processed foods like Bread, Chocolate etc, all have pesticide residues.

What is worthwhile to know is that 82% of 251 blood samples in a study conducted in Belgium (Charlier & Plomteux 2002) were found to have detectable amounts of pesticide particularly -p, p' – DDE (67% of samples). Urine was found to contain six pesticides or their metabolites in 50% of the samples of 978 adults participating in the Third National Health and Nutrition Examination in a study conducted in USA (Kieszak et al. 2002).

* The article is based on the report "The Minimisation of Pesticide Residues in Food: A Review of the Published Literature".

1. Summary

Current crop protection, against pests, weeds and diseases over much of world agriculture, is based on the use of pesticides. The scale of pesticide use inevitably results in the occurrence of residues in food products.

We have studied the publicly available scientific literature to identify the current scale of the problem and to identify a range of approaches to minimising both residues in individual foods and total exposure.

Our review of published information on pesticide residues indicates that a significant proportion of most major vegetable crops and all major fruit products consumed in the UK contain residues of pesticides. Similar situations exist in EU and USA. Residues are found both in produce which originates in UK and in that which is imported.

Fungicide and growth regulators are most commonly found as residues. Post harvest applications more commonly give rise to residues than do applications in the field. In current systems of production losses of food production would be large without the use of crop protection chemicals. Alternatives to chemical crop protection would need a substantial reconsideration of systems of crop production. Most current systems of production have come about because chemical crop protection has been available and so as to facilitate the use of pesticides. The concentration of research on systems designed to function with chemicals has meant that alternative options have not recently been explored. The pesticidal materials most commonly found in food products and the crops and foods in which residues are found most commonly are identified in the report.

Although residues are commonly found in most major food crops only a very small proportion of residues detected exceed Maximum Residue Levels (MRL), there is no substantive evidence that current residues represent a health issue. Although this was not an issue we were commissioned to examine. At the present time, in both EU and UK, the application of pesticides in agriculture is decreasing. A number of options which should further accelerate this decrease, and as a result reduce residues in food, are detailed here. These options are identified on the basis of our review of the publicly available literature.

There are also a number of options not related to current pesticide application practices and technologies, which could influence the quantity of residues or the frequency with which such residues are found. We have commented on these options.

Both biotechnology enhanced production systems and organic farming have as part of their basic rationale the reduction of pesticide use to low or no levels. Both approaches individually and as part of a whole UK strategy have a potential role to play in reducing residues in food.

Organic farming has been successful in reducing the frequency and levels of pesticide residues in produce from this sector. In many cases this means production of results in a yield penalty. The yield penalty seems more often to be related to the availability of soil nutrients that to the ineffectiveness of crop protection. Consequently many of the crop protection practices developed for organic farming seem likely to have value to conventional agriculture, especially Integrated Crop Management (ICM) systems. Introducing appropriate methods from this sector may therefore reduce in both pesticides use and recovery as residues. Many of the options available to conventional producers e.g. selection of varieties, timing of operations are also available to organic producers. They may result in improvements to total productivity so making organic production a more viable option for the production of some commodities. Pesticide residues currently detected in organic foods include a number e.g. dichloro-diphenyl trichloro-ethane (DDT) derived from historic uses in more conventional agriculture.

The use of Genetically Modified (GM) crops in crop production in USA has resulted in some reductions in pesticide use. Total reductions seem less than had been anticipated. For herbicide tolerant crops the ability of GM crops to reduce use depends on whether the broad spectrum herbicides to which the crops are now resistant can deliver a high enough standard of weed control to obviate the need for repeat applications or the continued use of residual materials. Varieties engineered to be resistant to insect pests, *bacillus thuringiensis* (Bt) varieties, have reduced the use of insecticides. Currently no commercial GM varieties seems to have been rendered resistant to fungal infection.

Pesticide use in conventional systems can be reduced by both advances in crop breeding and through modified agricultural practice. Plant breeding can reduce the need for fungicides and insecticides but is likely to have little impact on herbicide use other than through any impact of Herbicide tolerant (Ht) crops. Breeding for insect and disease resistance which has been successful in the past will continue to help reduce the need for fungicides and insecticides.

Our review of the literature indicates that modification to crop husbandry presents a wide range of means of reducing both pesticide use and residues. A large number of major system experiments, e.g. the Boxworth Project, have identified a range of techniques which have been effective in reducing chemical use. These system experiments have informed the production of Decision Support Systems (DSS) to provide a basis for rationalising pesticide use. Case studies detailing work on glasshouse salad crops, the Danish system, Glyphosate, Stobilurin fungicides and cereals are used in the report to illustrate some available options for reducing chemical use. Many of the suggestions made are of a type which would be applicable to both UK systems and countries outwith UK from where we currently receive imports. Options to reduce pesticide use include the possibilities of replacing chemical use with physical protectants such as mulches. Pesticide residues cannot be seen in isolation within the debate about food quality. Minimising pesticide residues through reduced use or alternative practices must be seen in the context of consumer demands for products with high visual impact.

A high proportion of the residues found in UK foods derive from post-harvest chemical use. A range of options related to both the use of crop protection chemicals and the management of the storage environment could reduce chemical use and the presence of residues. Options for fruit, grain and potatoes are detailed in the report.

Current levels of pesticide use and residues in foods are a function of the post-war development of farming systems based around pesticide use for crop protection. Moving away from current methods and levels of pesticide use will require some rethinking of the design of the current systems. A wider range of options for crop protection are needed. Such systems need to be financially and environmentally viable and deliver high quality food. The need to minimise pesticide residues should be seen in the context of the current environmental, agricultural and health debates.

2. Introduction

Pesticides are used in the food production process to control the effects of fungal diseases and invertebrate pests on crop development and production, to reduce the competition experienced by crops from other plant species (weeds) and to protect harvested food products from fungal and pest activity and premature development such as sprouting while in store. Pesticides became a novel agricultural practice, post 1945, increased in number and variety during the 1960s and became the standard practice in the 1970s. These changes and the widespread use of pesticides are documented through the addresses presented at the British Crop Protection Councils (BCPC) Annual Conference in the Bawden Memorial Lecture. The lectures given between 1973 and 1998 have recently been brought together into a single volume (Lewis, 1998). In the lectures given in 1974 and 1975 Sir Charles Pereira and Professor JM Hirst discussed the use of chemicals in crop protection but in the context of a range of other approaches. The 1977 lecture by Dr. J T Braunhotz, the then Director of ICIs Jealotts Hill Research Station, was wholly chemically based but the approach was subdued compared to the 1988 lecture by his successor Mr. J R Finney who was able to document the international agrochemical industry at its height. Crop protection with chemicals (pesticides) is now the norm for crop protection in both UK and world agriculture with around 350 active ingredients currently approved for use in UK and around 800 in one or more EU States. The development of crop protection in agriculture to its current level of dependence upon chemicals resulted in agricultural systems being developed so as to facilitate chemical management and has resulted in systems whose design assumes that crop protection will be effected with chemicals.

On a world scale an estimated 35% of potential crop production is lost to pests, pathogens and weeds each year. A further 20% is lost post-harvest. Examples of the impact of weed competition are given by Holm (1976). He showed that for a range of crops, including wheat, beans and potatoes that yield reductions due to weeds ranged from zero to 90% with means of 17-51%. Pesticide use, estimated at 2.5 million US tons per annum in 1992 (Pimentel, 1992), saves around 10% of world food supply.

Global arguments are of course of limited relevance to issues related to pesticide use in UK or pesticide residues in the UK diet except that the UK is a net importer of food, with much exotic produce as a constant feature of the British diet. Despite the above considerations Pimentel (1992), suggested that the damage caused by pesticides exceeds their benefit and promise. This contention, over a decade ago, resulted in the development of a questioning approach and a wider perspective on crop protection. The arguments advanced by Permentel (1992) have, principally in an environmental context, been expanded by Pretty *et al* (2000). They identified the total external costs of UK agriculture to be £2.3 billion yr^{-1}, £208 ha^{-1} of which the costs attributed to the presence of pesticides in drinking water were £120 mil yr^{-1}. In addition to direct effects related to water clean up pesticide manufacture contributes to those costs incurred with respect to air quality. It raises a number of clear issues for a small island country such as UK, where reduced chemical use in integrated pest management systems or in other systems, less dependent on pesticides, are real possible options.

The widespread use of pesticides has led to concern over pesticide residues in food being consistently reported by MORI and other surveys of public concern. An example of this concern is that over the period January 1999 to date "The Guardian" newspaper published 81 articles on the subject of pesticide residues. The majority of these focused on concerns.

Recent press coverage has suggested that around one third of fresh food products may contain significant pesticide residues and, for example, that 20% winter lettuces may contain residues above safety guidelines.

Pesticide residues may enter the human food chain directly or through the consumption of meat and milk products from animals that have consumed primary products containing pesticide residues. Cowie and Swinburne (1977) reviewed what was known at that time of the movement of pesticides into milk products. At that time the Organo-Chlorine (OC) insecticides were significant contaminants of milk as a result of their presence in the feed consumed by the animal. These materials are no longer in use in Europe, but they are still used elsewhere in the world and may occur in food products as a result of their long persistence in soil. Baker *et al* (2002) compared pesticide residues in fruit and vegetables originating from conventional, IPM and organic production systems. They found 23% of organic samples contained one or more residues, compared to 73% of conventional samples and 47% of IPM derived samples. Forty percent of the residues in organic samples were derived from banned OC compounds. After removal of OC materials, pesticide presence in organic vegetables fell to 9% of samples and, overall, to 13% of samples. Carrots, potatoes and other root crops, cucurbits and leafy greens are particularly prone to absorb OCs. On contaminated sites reduction of residues may require the selection of crops less likely to accumulate OCs from contaminated sites. This subject has been reviewed by Nash (1974), Maltina *et al* (2000) and Groth *et al* (2001). Herbicides such as 2, 4-D, 2, 4, 5-T and MCPA, which are still in use, have all been found in milk after they had been used to treat pasture or food subsequently fed to cattle (Cowie and Swinburne 1977). Despite this, however, the most significant route into the human diet is still through the direct consumption of primary products which have been treated with current materials. Current concerns have resulted in the UK, EU and USA and many other Countries setting up systems for the routine monitoring of food products, imported and indigineous. In the UK this role falls to the Pesticide Residues Committee (PRC) who provide data annually. The PRC data include a series of staple products e.g. potato and bread which are monitored at annual intervals and a range of other commodities monitored less frequently (PRC 2001, 2002). Results reported by the PRC for 2001 indicated that residues were present in 14% of 867 animal product samples compared to 40% of 551 cereal samples and 37% of 1963 samples of fruit and vegetables. Only 4 of 217 samples of milk contained residues, which compared favourably with residues in 55 of 144 samples of bread. Residues in primary products have therefore been targeted as the focus of this review although the residue of older materials, including residues in grassland, mean that such an approach will not achieve a complete absence of pesticides in the diet (Table 1).

The primary route through which most pesticides enter the food chain therefore is as a result of their application as crop protection materials in the production of combinable and other arable crops, vegetables and fruit crops and as a result of their application to fruit, vegetables and grain products during storage. The primary focus of this review of the published scientific literature is thus on the options currently available to reduce the need for pesticide applications within agricultural and horticultural food production and storage systems and the options for subsequently reducing the movement of applied pesticides into those parts of

Table 1: Residues Detected by Co-op in Samples Analysed in June 2001-August 2002

Substance	No. of Month/Crops Found	Main Crops	Type of Chemical
Heptachlor	1	Yam	Banned Insecticide
Methyl bromide	1	Lettuce	Banned Sterilant
Lindane	1	Mushroom	Banned Insecticide
Dieldrin	1	Passion Fruit	Banned Insecticide
Dithiocarbonate*	25	Fruits, Vegetables	Fungicide
Carbendazim*	19	Apple, Citrus	Fungicide
Iprodione	14	Vegetables	Fungicide
Thiabendazole	11	Apple, Citrus	Fungicide
Imazalil	8	Citrus	Fungicide
Captan*	7	Apple	Fungicide
Chlorpyifos	4	Fruit	Insecticide
Bromopropylate	4	Citrus	Insecticide
Metalaxyl	3	Apple	Fungicide
Diphenylamine	3	Apple	Fungicide
2 Phenylphenol	3	Citrus	Fungicide
Vinclozolin*	3	Vegetables	Fungicide

* on Co-op Restricted list.

agricultural products consumed as foods. The data of Baker *et al* (2002) mentioned above indicates that adjustments to crop production systems, through changes in crop species, crop varieties, crop rotations and management practices can all have significant impacts upon productivity and upon residues. They may also, however, influence quality attributes, such as the existence of bruises and availability at particular times. Crop protection may also influence the balance of field and protected production and the visible and nutritional qualities of food products. Actions to reduce pesticide levels in food through changes in pesticide use will therefore impact upon the environment. This is not considered in detail here, except where it is believed that a change in use, dedicated to reducing a pesticide residue in food, may have a negative environmental impact. The practicality of such changes and the public acceptability of products are both key issues. In its submission to the EU the European Crop Protection Association identified modified farming practice, and especially the use of Integrated Crop Management (ICM) systems, as one of the principal means of reducing potential risks associated with the current use of crop protection products.

Much of the food consumed in the UK comes from non-UK sources. Foods produced outwith the UK may show different occurrences of pesticide levels. Studies by the Pesticide Residue Committee for 2001 showed that 25% of the UK samples they tested contained detectable residues while 32% of imported samples contained residues. It is important therefore to review options for change in UK production systems in the context of the practices employed elsewhere. Within the general context identified above a series of key areas influence the role, scale and practicality of changes which can result from modifications to current practice.

The subject of this review, pesticide residues in foods (and their relationship) to pesticide application is one, which has developed a massive literature over the past forty years. In the review we have aimed to follow the FSA specification (Section 1) and so have concentrated on changes in production systems, and their design and management and especially on means of pesticide use, which might lead to reductions in residues. In identifying systems which might influence residues we have attempted to identify any major implications of such changes for the cost, availability or quality of food products. We have also reflected on major differences between UK and other systems of production, which appear to have influenced either pesticide residues or the use of pesticides. The impact of pesticides on health is a large and separate topic. It has received substantial recent treatment from studies such as those four Maroni and Tait (1993), Schulz *et al* (1990), Carpy *et al* (2000), Schneider and Dickert (1994) and Tuormaa (1995). Consequently this review does not consider the impact of pesticides or pesticide residues on health. Similarly the review does not consider the mechanisms used to establish current acceptable levels of pesticides in food, legislation, the potential costs or the environmental impact (except as discussed above) of suggested changes. All of these are beyond the scope of this study.

In the review we have not aimed to be encyclopaedic. We reviewed a large number of published accounts of work (over 1,500 papers were retrieved). We have however discussed and based a series of suggestions for change on what we believe are the most important sources together with a sample of recent papers which we think summarise substantive points or issues and are exemplars of the material available.

3. Pesticide Use and Pesticide Residues

The starting base of any study on the reduction of pesticide residues must be information on current pesticide use, why pesticides are used and what residues are currently being detected in the foods available to UK consumers. In this section we therefore review the impact of disease on crops (fungicides represent the residues found most commonly in foods; Table 1) and the impact of weeds on crops (herbicides are the most commonly used pesticides; Table 2). This leads to consideration of current usages of these and other materials and the mechanisms by which they may move to food products. Not all pesticide applications lead to residues and so we consider residues found in foods both in government and supermarket studies. This leads to a series of provisional thoughts on consequences of reducing or even ending pesticide use, which then provide a starting point for the remainder of the review.

3.1 Why Pesticides are Used in UK/EU Agriculture and Horticulture

The term pesticide was formally defined for the purposes of the Food & Environment Protection Act 1985 and the Control of Pesticides Regulations 1986 to include any chemical or biological product use to kill pests or to prevent their damage or unwanted effects. The term pest was effectively widened beyond insects and other animals to include fungi, algae, and weeds. Animal repellents and plant growth regulators were also included within the term pesticide. Microbial products were classed as pesticides whereas parasitoid insects and predatory mites, for example, were excluded. The use of products marketed for other purposes, e.g. soaps, was outlawed unless appropriately registered. The Food & Environment Protection Act empowered the Pesticides Registration Directorate (PSD) to register all pesticidal products and inevitably brought their continued use under scrutiny.

Three hundred and eighty five active ingredients were listed in the first register of pesticides approved for use in agriculture and horticulture (MAFF/HSE, 1986). The intervening 16 years have seen a net loss of 51 active ingredients for use in agriculture and horticulture in the UK (PSD, 2002). The pressure for retention of these products for both the commerçial and the garden products market has

Table 2: The Pesticides Detected Most Often in the January-June 2001 Survey Carried out for the Pesticide Residues Committee

Substance	No. of Samples Assessed	No. with Residue (%)	No. > MRL	Compound Type
Thiabendazole	209	59 (28)	-	Fungicide
Imazalil	231	79 (34)	-	Fungicide
Carbendazim	210	45 (21)	5	Fungicide
Iprodione	151	45 (30)	-	Fungicide
Chlormequat	113	44 (39)	1	Growth Regulator
Chlorpropham	103	36 (35)	-	Growth Regulator
Captan	138	27 (20)	-	Fungicide
Dithiocarbamate	202	27 (13)	-	Fungicide
Chlorpyrifos	98	26 (27)	1	Insecticide
Methidathion	107	20	-	Insecticide
2 Phenyl phenol	50	18 (36)	-	Fungicide
Tolylfluanid	69	12 (17)	-	Fungicide
Carbaryl	81	11 (14)	-	Insecticide
Dicofol	81	9 (11)	1	Insecticide
24 D	33	9 (27)	-	Herbicide
Oxadixyl	99	9 (9)	-	Fungicide

been sustained, with particular emphasis on herbicides and fungicides; the market for insecticides is in decline. The total UK market for chemicals fluctuates annually as a result of variation in weather; generally heat encourages outbreaks of insects and insect-borne viruses while moisture promotes weed growth and many fungal diseases. The recent absence of severe winters has increased overwintered pests, pathogens and volunteer weeds, such as potatoes.

Mainstream UK agriculture and horticulture have become dependent on the use of pesticides since their discovery and invention, mainly as arsenical and copper-based products, at the end of the 19th Century. The knowledge that pesticides can be used either prophylactically or to repress an outbreak has resulted in the strict seasonality of cropping being replaced by use of any window of opportunity to sow crops regardless of the heightened pest and disease pressure that early-sown crops experience. The success of continuous successional cereal crops and autumn-sown cereals is at least in part due to the availability of cost-effective "insurance treatments" to prevent pest and disease attack when the crop is stressed in cool conditions. Up to the present time chemistry has kept pace with changes in pests and diseases. The withdrawal of the persistent organochlorine insecticides coincided with the discovery of synthetic pyrethroids with toxicity to pests at much lower doses than any other type of pesticide.

Public perception of the insidious nature of synthetic pesticides began with direct poisoning of operators by DNOC in 1947. The response to this was the Agriculture (Poisonous Substances) Act 1952, based on the recommendations of the Zuckerman Working Party (Conway *et al.*, 1988). Introduction of many new types of pesticides raised the first fears about pesticide residues in food, resulting in the formation, in 1957, of the Pesticides Safety Precautions Scheme (PSPS), a mandatory system for clearing the use of pesticides on the basis of their safety. This worked in conjunction with the Advisory Committee on Pesticides (ACP). Concern about wildlife resulted in a Wildlife Panel being added to PSPS in 1962 coincident with publication of "Silent Spring" (Carson, 1962). In 1961 the House of Commons Select Committee on Estimates caused the withdrawal of certain organochlorine insecticide seed dressings because of the death of grain-eating birds in 1957. Concerns about operator safety, food contamination and wildlife have therefore resulted in government actions since the early 1950s. Pesticide usage has however continued because of their immediate cost-effectiveness and support for prevailing crop systems.

The development of the use of insecticides, fungicides and herbicides has been reviewed on many occasions. Pesticide use is important because current systems of farming have been designed around the use of pesticides. Current farm sizes, structures and uses of labour are, at least in part a product of pesticide use. Changes in pesticide use thus need to consider social and economic issues as well as alternative means of pest, weed and disease control. Wider issues of this type are discussed in Section 8 of this review. At a time of considerable financial problems in farming these wider issues are of considerable importance.

The European Commission publishes annually a compilation of the national situations in the Member States and those European Free Trade Association states that have signed the Agreement on the European Economic Area. In 2000 around 45,000 samples were analysed for, on average, 151 pesticides (European Commission, 2002). Pesticide residues were detected at or below the MRL in 35% of cases. Only 4.3% were above national or EC-MRLs. The increase in samples exceeding MRL over earlier years (3.3% exceedance in 1999 and 3.1% in 1998—see website for European Commission, 2002) was put down to (a) changes in MRLs (set to lower limits), (b) more sensitive analytical methods (presumably then related to MRLs derived from obsolete techniques), (c) a broader spectrum of analyses being sought, and (d) better information flow with the European Union, (sampling priorities can change overall in the event of the detection of a problem in one member state).

In 2000 fungicides continued to dominate the findings (European Commission, 2002). A change in reporting, from absolute numbers of findings (1996-1999) to relative frequency of pesticide occurrences, makes it difficult to compare year-on-year. The 2000 findings were affected by many single residue methods targeted towards specific problems. Of the twenty pesticides surveyed, on a co-ordinated basis, residues of the maneb group were found most often (16%), followed by vinclozolin (4%) and benomyl (2%). Residue exceedance was greatest for the maneb group, on cabbage, cucumbers, peas and rice. Assessment of chronic exposure indicated that Accepted Daily Intake (ADI) were not exceeded in these commodities, but a quasi-probabilistic analysis of acute exposure to methamidiphos on cucumbers indicated that Acute Reference Dose (ARfD) would have been exceeded for toddlers (this ARfD is yet to be fully established).

In establishing what pesticides are currently found in the food products consumed in UK, both home grown and imported produce, is a significant component of this review. The presence of residues, which must result from the application of pesticides during production or storage, must however be put in the context of the impact of disease on crops. On the basis that fungicides are the compounds most frequently found as residues in the UK, by supermarket groups and government (Tables 1 and 2) and in types of pesticides in the USA (Table 3) and that herbicides are the most widely used pesticides much of our subsequent emphasise is placed upon these compounds (Table 4).

Table 3: The Residues Found Most Often in USDA Studies of Pesticide Residues 1992

Commodity	Pesticide	Type of Material	Frequency Found (%)
Oranges	Thiabendazole	Fungicide	64
Potato	Chlorpropham	Growth Regulator	59
Apple	Thiabendazole	Fungicide	57
Peach	Iprodione	Fungicide	54
Grapefruit	Thiabendazole	Fungicide	54
Peach	Dicloran	Fungicide	47
Celery	Permethrin	Insecticide	39
Banana	Thiabendazole	Fungicide	38
Celery	Chlorothalonil	Fungicide	32
Apple	Azinphos-methyl	Insecticide	31

Comparison of the residues, in fresh produce, most commonly detected by the Co-op (this is the supermarket whose residue data is most easily accessed), by the UK Government and the USDA (Tables 1, 2 and 3) shows large similarities. A small number of materials e.g. Thiabendazole are responsible for a disproportionately high proportion of product/residue samples. This suggests that many of the problems which pesticides are used to ameliorate are ubiquitous on an international scale and that solutions found to be of value in UK may well be applicable outwith the UK and so impact on imported produce.

3.2 Impact of Disease on Major Crops

Crop diseases caused by fungi can lead to high, sometimes complete losses of crops. This is most devastating where the epidemic is spread over large areas and if all plants are more or less susceptible to the pathogens. Fungal disease can results in not only a quantitative yield loss but also losses in quality due to lower food quality and to decreased potential for storage and mycotoxin contamination. Most crops are attacked by several pathogens each adapted to certain conditions such as the downy mildews and fruit rots that enjoy moist conditions, others need only very limited moisture such as the rust fungi. Besides leaf pathogens there are also soil borne pathogens.

There are many historical examples of crop losses from disease that have had significant impacts on food supply. Late blight of potatoes caused by *Phytophthora*

Table 4: Sales (£M, not corrected for inflation) of commercial agricultural and horticultural pesticides in the United Kingdom by members of the Crop Protection Association (formerly the British Agrochemicals Association), 1985-1999. Sales of seed treatments are not included in the Table: they ran at £M11-25 over the same period.

	Herbicides	Insecticides & molluscicides	Fungicides
1999	186	38	139
1998	182	42	147
1997	204	39	141
1996	227	45	147
1995	214	50	139
1994	195	49	116
1993	178	49	124
1992	162	40	131
1991	187	37	128
1990	180	40	128
1989	191	47	133
1988	202	37	109
1987	178	30	97
1986	176	25	95
1985	174	29	98

infestans led to periods of starvation in Ireland in the 19th Century and caused problems in food supply in Germany in 1916 and 1917. The detrimental effect of plant disease stimulated the development of fungicides, the oldest of which was Bordeaux mixture on grapes. Early fungicides were very broad spectrum and included products such as the organotins. The first systemic products were introduced in the 1960s, which were more specific, and safer to the crop.

The impact of plant diseases and the most important plant pathogens can be differentiated by crop and the crop losses they cause. The loss potential of diseases is higher in areas of high input cultivation than in other regions. In cereals intensive disease control practices reduce losses by almost two thirds from a potential loss to disease of 20% down to an actual loss of 7% (Table 5).

Table 5: Loss Potential and Actual Losses Due to Diseases 1991-1993 (Dehne and Oerke, 1998)

	World wide		Western Europe	
Main Crops	Potential Losses to Disease	Actual Losses to Disease	Potential Losses to Disease	Actual losses to Disease
Soybean	11%	9%		
Oilseed rape			11%	5%
Sugar-beet			15%	7%
Coffee	26%	14%		
Wheat	16%	11%	20%	7%
Rice	21%	12%	21%	9%
Maize	10%	10%	4%	5%
Potatoes	25%*	17%	25%*	10%

* 25% figure for actual losses in parts of South America and Africa where very limited crop protection measures are taken. In more intensive situations, as are practised in the UK and in countries supplying the UK potato market, potential losses are much higher.

Fruit, particularly citrus fruit account for a disproportionate number of residue problems and together with vegetables account for 40% of the total fungicide use worldwide (Table 6).

Table 6: Summary of the Most Important Diseases in Fruits – Citrus, Pome Fruits, Stone Fruits and Grapes (Oerke *et al*, 1994)

Citrus	Pome fruits	Stone fruits	Grapes	Bananas	Soft fruits
Phytophthora diseases	Scab	Brown rot	Powdery mildew	Black sigatoka	Grey mould
Scab	Powdery mildew	Cherry leaf spot	Downy mildew	Brown sigatoka	Phytophthora diseases
Black spot	Crown and root rot	Leaf curl	Grey mould	Panama disease	Leaf scorch and Leaf spot
Melanose	Canker	Root and crown rot	Black rot		Fruit rot
Mould diseases	Mould diseases	Powdery mildew	Eutypa dieback		Mucor spp

The most significant loss causing diseases in vegetables are summarised in Table 7.

Table 7: Summary of the Most Important Diseases in Cucurbits, Tomato, Cabbage and Other Vegetable Crops (Oerke *et al*, 1994).

Cucurbits	Tomato	Cabbage	Beans	Others
Powdery mildew	Late blight	Damping-off	Black root rot	Damping off
Black root rot	Early blight	Downy mildew	Rhizoctonia	Downy mildew
Downy mildew	Wilt diseases	Blackleg, canker	Angular leaf spot	Leaf spots
Damping off, root rot	Grey mould	Dark leaf spot	Rust	Grey mould
Angular leaf spot	Corky root rot	Black rot	Bacterial blight	Powdery mildew

3.2.1 Approaches to Control

The control of disease is commonly managed by integrated control measures including crop rotations, use of resistant cultivars, proper soil preparation and appropriate fertiliser application. All of these preventative measures can be optimised to reduce inoculum density. Plants are particularly vulnerable during seedling emergence and the use of seed treatments are often necessary to safeguard juvenile stages when significant plant losses can lead to total crop failure. Foliar sprays predominate in the application of fungicides. The number of foliar applications range, depending on crop and target pathogen from one per year to several treatments in spray schedules (Schwinn, 1992).

3.3 The Impact of Weeds on Crops

Much of the early research on herbicides was related to characterising the impact of weeds on crop growth. Aldrich *et al* (1975) assessed the effect of the duration of weed control on maize yields. Allowing weeds to grow for two weeks reduced yield by 8% but 5 weeks weed growth reduced yield by 17%. Similarly, Walker *et al* (1984) found that weed density had a curvilinear relationship with soybean yield with high densities reducing yields by 80%. Major effects on cereal yields in UK are caused by both grass weeds such as *Alopecurus myosuroids* and broad-leaved weeds such as charlock (Wilson *et al*, 1985) with yield depression being a function of the duration of competition. In fruit plantations data reviewed by Atkinson and White (1980) showed that weed competition could reduce growth by 15-96%.

Effects on fruit yields in cropping trees could be as high as 35% with effects on financial returns, as a consequence of effects on fruit grade out, higher at 45%. In stone fruit growth reductions due to weed competition could be 50% (Rupp and Anderson, 1980). The size of the impact of weeds on crop growth and productivity, together with effects on the ease of harvesting and the impact of pests and diseases has resulted in herbicides being the most commonly used pesticide (Table 4).

3.4 Current Use and Trends in Pesticide Use

The overall usage of pesticides in the UK is almost certainly past its peak (Table 4). Sales by members of the Crop Protection Association (Trow-Smith, pers. comm.) began their decline in the mid-90s and are likely to be reduced further either as a result of the success of the Voluntary Initiative, or certainly in the event of its failure.

Ewald and Aebischer (2000) reviewed trends in pesticide use in England between 1970 and 1995. In general the area treated and the intensity of use increased over the period of study. The spectrum of activity of pesticides applied to arable crops increased from 22 susceptible taxa in 1970 to 38 in 1995. Use of pesticides and fungicides in break crops and insecticide use in spring crops was less than average use. Monoculture wheat received similar pesticide treatments but more fungicides and insecticides than traditional rotations.

Information is available from government surveys of pesticide use, which also comment on sectoral trends. Kerr and Snowden (2001) have documented current pesticide use in arable crops in Scotland. This showed that compared to their previous survey in 1998 the area of arable crops had fallen by 8%. Insecticide applications were highest in seed potatoes (>80%) but under 20% for many cereal crops. Fungicide and herbicide use was maintained. Around 100% of cereal, oilseed rape (OSR) and potato crops were treated with both herbicides and fungicides. In cereal crops the fungicide treated area remained at the 1998 level while the herbicide treated area increased. UK figures are shown in Figure 1.

Fungicides are used extensively in UK agriculture, and throughout the world, to maintain the yield, quality and appearance of produce. They can be applied at any time from pre-sowing seed treatments through to post harvest applications direct to edible produce.

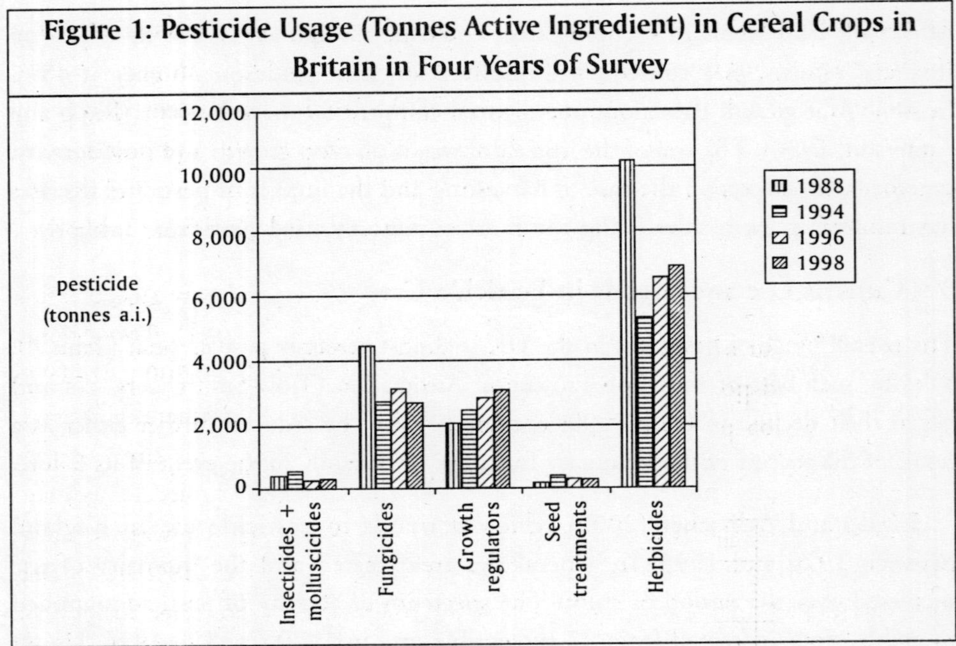

Figure 1: Pesticide Usage (Tonnes Active Ingredient) in Cereal Crops in Britain in Four Years of Survey

World-wide trends in international pesticide use in relation to perceived risks have been discussed by Ishii (2000) and trends in China by Zahang *et al* (1997). Decreasing use seems to be a common factor in most sectors. Ward and Armstrong (1998) document decreasing use of pesticides in sheep production in Australia. In the period 1964-1982 pesticide use on major crops in USA more than doubled (Lin *et al* 1995) increasing from 106 to 278 Gg Active Ingredient (AI). Since this time, however, use has declined to 261 Gg ai in 1992. Similar trends were also reported by Foy (1993), Osteen (1992) and Dexter (1991). Osteen and Szmedra (1989) attributed the earlier increases to increased herbicide use especially on maize and soybean and the latter decreases to regulatory decisions removing older chemicals from the market. Swanson and Dahl (1989) however attributed a 33% reduction in pesticide use between 1980 and 1986 to a combination of a depressed farm economy and then discriminating use of more active products. The overall decrease in pesticide use has concealed a rise in herbicide sales.

Trends in other parts of the world have not always paralleled those in the USA and EU. In Bangladesh pesticide use increased in the late 1980s. Here 95% of pesticide use involved insecticides with organophosphorus compounds making up over 60% of this group. Soerjani (1988) discussed pesticide use in Asian countries including India, Korea, Indonesia, Malaysia and Pakistan and the need to balance use with risks. Trends throughout the tropics have been detailed by Delp (1986), while Bellotti *et al* (1990) and Constenta (1988) have documented the position in Latin America. Bellotti *et al* (1990) found that pesticide use was continuing to grow as a result of moves to more mechanised monoculture enterprises.

In addition to the above trends a significant change in industry structure has been the growth of the generic sector (Sissan and Williams, 2001). The generics market is predicted to grow from $20b in 1997 to $27.5b in 2007. In 1996 patent protected compounds accounted for 47% of the world agrochemical market. It is anticipated that by 2005 this will have fallen to 30%. This suggests that future chemical crop protection will be based on a series of older molecules, perhaps with residues of greater consequence. In 1997 the global market was valued at $33.5b and growing at 1.5-5% annually. This market comprised 48% herbicides, 27% insecticides and 20% fungicides. In 2005 the market is projected to be 59% herbicide, 22% fungicide and 19% insecticides. Glyphosate which currently enjoys sales of $2b is projected to increase in use, aided by Roundup Ready Crops. These projected changes are important against a background where herbicide residues in foods are rare and fungicide most common. Trends in the use of herbicides over the period 1988 to 1998 have been summarised by Orson and Thomas (2001).

3.4.1 Soft Fruit

A 38% decrease in soft fruit area in Great Britain from 1990 to 1998 (9,430 ha) was associated with a 49% decrease in insecticide usage by weight, the equivalent figures for acaricides, fungicides and herbicides being 83%, 39% and 39% respectively (Garthwaite & Thomas, 2000). The large change in acaricide use was associated with replacement of endosulfan on blackcurrants with fenpropathrin, effective at a lower dosage, and with increased usage (808 ha in 1998) of the predatory mite *Phytoseiulus persimilis* Athias-Henriot for control of the two-spotted spider mite (*Tetranychus urticae* Koch). The heaviest use on soft fruit crops is, decreasingly, fungicides and, increasingly, soil sterilants (Table 8).

Table 8: Pesticide Usage on Soft Fruit in Great Britain (Garthwaite & Thomas, 2000) *(tonnes)*

	Area (ha)	Acaricides	Insecticides	Fungicides	Herbicides	Soil Sterilants
1990	15,102	103	11.2	116.7	55.3	51.3
1994	12,520	4.6	7.5	107.9	59.0	147.3
1998	9,430	1.7	5.7	80.0	33.8	197.5

3.5 Pesticide Movement to Food Products

3.5.1 Pesticide Metabolism

Pesticides are incorporated into plants through metabolic processes. They are rarely eliminated unaltered via roots or leaves except when shed in dead tissue. Their metabolism may be divided into three phases (Schmidt, 1999). Phase I comprises transformation by oxidation, reduction or hydrolysis into primary metabolites, or exocons. In phase II the exocons conjugate by means of –OH, –SH, NH_2, -COOH, –CL or $–NO_2$ groups to endogenous substances, endocons, which tend to be more hydrophilic than the original pesticides or its exocons. Endocons can be sugars, amino acids, organic acids or glutathione. Phases I and II are mediated by enzymes. The bound substances are then internally secreted, mainly into cell vacuoles, in Phase III. Plant metabolism studies with new pesticides focus on Phases I and II, but the bound substances remain undetected, except when the plant tissue is destroyed by combustion. Less radical treatment, using acid, alkali, enzymes or solvent/water extraction, reveals a number of residue products, their extent dependent on the nature of the original plant material. Cell wall components, lignin, pectin, cellulose, hemicelluloses and protein will bind C, with lignin, which provides the major excretory route. In this way C in labelled methyl bromide can find its way into seed DNA (Mostafa *et al.* 1992). The point at which a breakdown process results in the transformation of pesticide-derived material into a harmless endogenous substance is clearly a key issue. However aromatic and heterocyclic ring structures remain largely intact within bound substances extractable only by combustion. Aliphatic chemicals and aliphatic side chains are largely recycled as carbon skeletons. In addition some pesticides, such as phenoxyacetic acids, enter into a reversible sorption with cutin polymer immediately upon arrival on the leaf's surface (Tam *et al.*, 1996). The same problems may apply to a greater extent with some naturally occurring pesticides, which are nevertheless xenobiotic within most plants. Avermectin B_{1a} is an insecticide produced by the actinomycete *Streptomyces avermitilis*. Extraction from celery leaves using hot dimethyl sulphoxide, which solubilises lignin, found 19 and 15% of ^{14}C- or tritium-labelled avermectin respectively. The corresponding values for the stalks were 9 and 10% (Feely and Wislocki 1991). This non-extractable residue was found to have been metabolised and incorporated into glucose. This binding process is important to both the interpretation of residues in foods and an understanding of the degree of control which can be effected through Good Agricultural Practice (GAP) and other means of regulation (Figure 2).

Figure 2

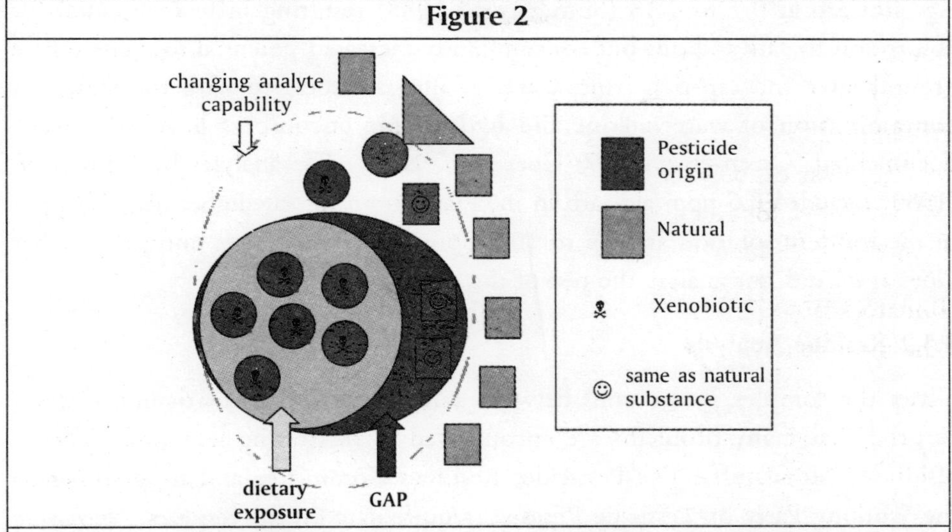

Analysis of xenobiotics showing the difference between what is required for dietary risk assessment and what would ideally be used to verify GAP. Analysts strive to extend the capability to include other pesticide metabolites, some of which may be bound to plant substances and rendered non-toxic, others, which are unrecognisable as being the same as natural plant substances.

Note: That this oversimplification ignores the possibility that some pesticides may cause plants to release substance toxic both to the target and to the consumer.

In addition to direct reactions within the plant pesticides are subject to other forms of degradation. Photolysis, photochemical transformation by solar radiation varies greatly between pesticides (Vialaton *et al.*, 2001) as might be expected, given that photostability is a desirable attribute for a foliar pesticide. Ozone denatures organic pesticides by various oxidation processes, also by dealkylation, hydrolysis and dehalogenation, but leaving recognisable residue products, and having a limited effect on fully substituted benzene ring structures (Ohashi *et al.*, 1993). Unnatural processes, such as irradiation (Lepine, 1991) may enhance this process in foodstuffs treated outwith the UK.

The systemicity of a pesticide is often crucial for its success, and will affect its distribution within the plant. Thus the systemic insecticide dimethoate could be detected in orange pulp when six other organophosphorus insecticides, all with contact action, could only be found in peel (Cabras *et al.* 1995). Systemic compounds may be expected to have low octanol-water partition coefficients, e.g. aldicarb at 0.7 to 1.13 (Suntio *et al.* 1988) resulting in low potential for absorption by fats and oils but concomitantly increased potential to contaminate groundwater and sap-rich fruit. Cases of illness associated with the aldicarb's contamination of watermelons and hydroponic cucumbers have been well-documented (Green *et al.* 1987; Goes *et al.* 1980). The analyses by Goes *et al*, (1980) included 0.6 ppm aldicarb in the gravel from the cucumber bed, 1.8 ppm in the nutrient solution and up to 10.7 ppm in the fruit. It is unfortunate that Goes *et al.* did not analyse the rest of the plant.

3.5.2 Residue Analysis

Given the complex interactions between xenobiotics and plant tissues, it is no surprise that many problems are encountered in registering levels of pesticide residues in foodstuffs. The Pesticides Residues Committee and its predecessor the Working Party on Pesticide Residues require data on the residues, *"including any specified derivatives such as degradation and conversion products, metabolites and impurities which are considered to be of toxicological significance"* (Working Party on Pesticide Residues, 2000). Analysis follows *Codex Alimentarius Commission* guidelines. There are significant concerns about changes in detectable pesticide levels dependent on the state and type of the material being processed. There was, for example, a greater apparent loss of pesticides from lettuce, oranges and tomatoes when processed at room temperature as opposed to when part-thawed, the opposite "*inexplicably*" (op. cit., p. 13) being the case for apples. There was less loss in tomatoes and oranges than with lettuce or apples. These differences may reflect variation in availability of endocons and the stability of the endoconexocon bond. It is claimed that cryogenic milling (grinding of pre-frozen samples with dry ice) minimises apparent losses.

Analysis of pesticide requires several steps, extraction from the matrix, separation of the analyte from other material extracted in the first step, detection and quantification. Traditionally, pesticides have been extracted from aqueous material using dichloromethane and from plant and soil material using organic solvents or inorganic acids and alkalis. Separation of volatile analyses is usually by capillary column-gas chromatography (GC) associated with electron capture, nitrogen-phosphorus or mass spectrometer detectors. Non-volatile or heat-sensitive analyses are usually separated and detected by high-performance liquid chromatography (HPLC) associated with an ultraviolet diode array detector. The latter is, however, inadequate for traces of polar or thermally unstable analyses, and HPLC is in any case less efficient than GC separation. Thus there will be different problems associated with a range of metabolites arising from the use of one pesticide, let alone differences in efficiency over the range of pesticide products. The potential of newer methods of extraction has recently been reviewed in the context of extraction from humic materials (Koskinen and Duffy 2001).

The residues of some pesticides currently cannot be measured reliably while others require special methods. For example, glufosinate could not be determined in potatoes in the 1998/1999 surveys. Dithiocarbamate residues are measured by generation of carbon disulphide from ungrounded plant portions, but this cannot be done for brassica and watercress, which generate carbon disulphide from natural substances. It is probable that difference in Chlorothalonil residues between lettuce and onions relates to sulphur compounds in onions, this problem being resolved by processing onions under highly acidic conditions (Working Party on Pesticide Residues, 2000). The complexities of residues are such that for some materials the amounts detected will be influenced by advances in extraction technologies while the proportion of samples containing residues will be influenced by developments analytically technology.

3.6 Human Exposure via Food as Opposed to Other Sources

In a study concerned with potential exposure to carcinogens in the USA (Dougherty *et al.*, 2000), benchmark concentrations were exceeded for six pollutants including chlordane, DDT and dieldrin, by the age of 12. Food consumption was the most important route for exposure, even to industrial pollutants. Exposure through the consumption of fish accounted for a large proportion of food exposures.

A survey of semen quality of farmers in Jutland before and after the main period of pesticide spraying (Juhler *et al.* 1999) was intended to compare those engaged in orthodox as opposed to organic production. The estimated dietary intake of 40 pesticides did not entail a risk of impaired semen quality, but the study highlighted differences in ADI between farmers when divided into three groups with no, less than half and more than half of their diet being organically produced fruit and vegetables. The estimates of pesticide intake were significantly lower in the latter group, but all groups were at or below 1% of ADIs, except for dithiocarbamates (maximally 0.21 microg/kg day, 2.2% ADI), methidathion, (maximally 0.01 microg/kg day, 1.4% ADI), and 2-phenylphenol (maximally 0.21 microg/kg day, 1.1% ADI). Occupational exposure to pesticides seems to be of less importance than dietary intake.

Some pesticides are used in a variety of situations, e.g. until recently, the insecticide dichlorvos in domestic situations, in food storage and to control fish lice on farmed salmon. Carbaryl has been used as a soil and foliar insecticide, but persists on the UK market as a fruit-thinning agent for apples. Pentachlorophenol is a persistent biocide, used in some parts of the world for the control of wood-boring insects and fungal rots, for total weed control and as a post-harvest defoliant and general disinfectant. Only the oldest of the synthetic insecticides, DDT, has a degradation process understood sufficiently well to provide guidance on its origin, with o, p'-DDT indicating very recent use, p,p'-DDT indicating exposure to recent use, p,p'-TDE indicating metabolism via animals or microorganisms, and a predominance of p,p'-DDE indicating historic use (Working Party on Pesticide Residues, 2000). The fungicide carbendazim (MBC), is produced by the MBC-generator fungicides, i.e. by benomyl quickly and thiophanate-methyl slowly.

Some potentially illegal uses of pesticides have been found to result from legal applications to neighbouring crops. Mushrooms were recently found to be contaminated with chlormequat used on the cereal straw on which they were cultivated. (Table 2).

It is clear from the literature reviewed above and studies such as that or Maroni and Tait (1993) that ingestion via the diet is the principal means of exposure.

3.7 Pesticide Residues Found in Humans

The study of pesticide residues in humans has focussed almost exclusively on the persistent organochlorine insecticides, often in association with pollutants of a different origin, in particular polychlorobiphenyl compounds (PCBs) and inorganic toxins. The main residues tested are DDE (from DDT) and HEOD (from aldrin and dieldrin).

Bates (1990) describes the definition of the Maximum Residue Limit (MRL) which was introduced by the Codex Committee at its 1989 meeting. The definition includes a definition of Good Agricultural Practice (GAP) which is discussed later in the review. The definition includes GAP under the actual conditions needed for effective pest control and to encompass a range of levels of pesticide application in such a way as to leave the smallest practical residue.

In addition to focusing risk on the infant, breast milk provides a non-invasive medium for studying strongly lipophilic contaminants and has been surveyed for pesticide contamination since the 1950s (Harris *et al.,* 2002), with significant reductions in organochlorine residues from the 1960s onwards (Collins *et al.*, 1982). Fat content, typically 3.5%, varies considerably, and this has greatly affected the conclusions from surveys, in particular the associated risk assessments. However, long term studies of breast milk contamination provide an indicator of environmental contamination, and can serve to focus on the risks attached to particular diets, e.g. studies in Beirut gave a positive correlation between residue levels and a diet dominated by high fat meats and tuna, a weak correlation with poultry consumption, and a negative correlation with vegetable oils (Dagher *et al.,* 1999).

Blood sampling is not frequently reported in the literature, and most reports still focus on organochlorine pesticides. For example, 82% of 251 blood samples in Belgium (Charlier & Plomteux 2002) were found to have detectable amounts of pesticide, in particular -p,p'-DDE (67% of samples). It is clear that for most of the population exposure through food is the main route of pesticide exposure and entry.

Urine was found to contain six pesticides or their metabolites in 50% of the samples of 978 adults participating in the Third National Health and Nutrition

Examination in the USA (Kieszak *et al.* 2002). Three of the compounds—1-naphthol, pentachlorophenol and 3,5,6-trichloro-2-pyridinol—were possibly related to food consumption but no relationship could be established between reported food consumption and the compounds present. However, those who reported recently being exposed to pesticides in garden or household use did have higher levels of pesticides and their metabolites in their urine than those reporting no such exposure.

Current techniques used to access consumer exposure to pesticides are currently being reviewed. Modified methods make greater use of probabilistic approaches and models. Progress remains limited by a lack of sufficient data to allow quantitative analysis. The current approach to dietary risk assessment is based on the production of estimates which can be compared with actual daily intakes (ADI). Uncertainty and variability information on consumer exposure assessments is needed to improve estimates (Ferrier *et al*, 2002). The basis of a risk characterisation process is shown in Figure 3.

Figure 3: The Risk Characterisation Process

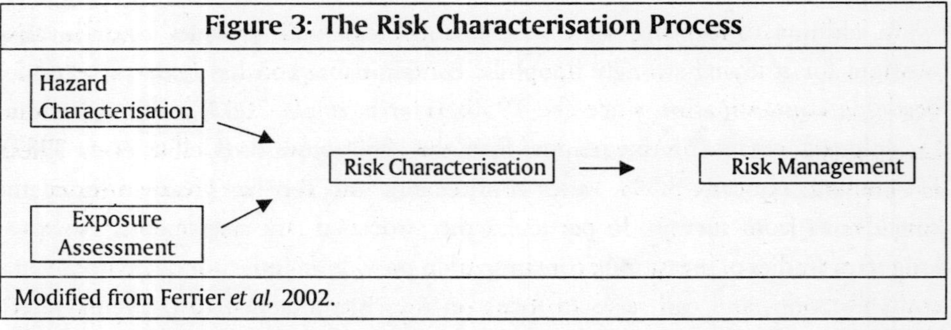

Modified from Ferrier *et al*, 2002.

In the context of this review perceived hazards usually result from the materials selected for crop protection and the rates at which they are used. Risks can be reduced by the selection of a less hazardous material or by the reduction of the rate of application. In general terms this usually reduces the residue in the food product. Exposure can be reduced by modifying the means of use and the material and the quantity of material, which ends in the food product. The probabilistic approach suggested by Ferrier *et al* 2002 would make it easier to use information of the above type than is the case with current deterministic models. Renwick (2002) reviewed the relationship of pesticide residues to hazard characterisation and intake estimations. He concluded that a disfunctionality was created as a result of MRLs being based on GAP but subject to a MRL not exceeding actual daily intake (ADI) or acute reference dose (ARfD). Intakes can be substantially below ADI. This disconnection, while delivering significant safety, gives a lack of transparency and makes it more difficult to link agricultural practice and ADI in a way which allows production to be managed to deliver low residue levels.

3.8 Current Residue Levels in Foods

Information on the scale of residue problems and in relation to particular chemical materials and crops are detailed in Tables 9, 10, 11, 12.

The principal current sources of data within the UK are the surveillance reports of the Pesticide Residues Committee (PRC), established in 2000 to replace the Working Party on Pesticide Residues (WPPR). Other sources of data, not necessarily substantiated by quality assurance data, are generated by the food industry and consumers' associations, with some results being published. The PSD operates an enforcement programme. PRC produces both quarterly and annual reports, the latest being for 2001 (PRC, 2002). WPPR produced annual reports as supplements to *The Pesticides Monitor*. PRC oversees the monitoring of the UK's food and drink in a three part programme of checks:- that no unexpected residues occur; that residues do not exceed Maximum Residue Levels (MRLs); and that human dietary intakes are within acceptable levels. These acceptable levels derive from levels of exposure primarily established in short and longer term studies of pesticides in mammals:

Residue levels in individual crops suggest approaches, which are likely to minimise residues. Data from the Working Party on Pesticide Residues for years before 2000 and now from the Pesticide Residue Committee provide basic information on the scale of the issue. This is indicated in Table 9. The change in the % of bread samples containing residues between 1999 and 2000 shows the impact of the inclusion of new residues in those sought and advances in analytical technology. The presence of residues in potato remains high. Currently around a quarter of samples measured for PRC contained detectable pesticide residues. More detailed information on the period Jan-June 2002 (obtained from the Pesticide Residue Committee website) is summarised in Table 10. This emphasises that although significant numbers of commodities contained residues few of these exceeded the MRL. The greatest proportion of materials found were fungicides.

A significant proportion of samples contained more than one residue. Fruit and vegetables were the main source of residues with the highest number in crops currently stored prior to sale or use. A rather small number of individual pesticides (Table 2) accounted for most residues found. Similar residues were found by the Pesticide Residues Committee (Table 2) and the Co-op, one of the few stores to list its analytical findings on the Web (Table 1). Co-op analyses picked up a number of older, banned pesticides (Table 1) which did not seem to be detected in Pesticide Residue Committee analysis. Data for produce measured in different months (Table 11) shows significant variation between different periods of the year when pesticides are most commonly found. The high levels found in winter months may relate to the impact of stored and imported produce. Similar analyses are known to be undertaken by other supermarkets but their data is not currently publicly available.

Table 9: Pesticide Residues in UK Food Samples

	1999	2000	2001
No. of Samples Analysed	2300	2304	4006
% with measurable residues	27	28	29
% with residues >MRL	1.6	1	0.7
% of sample of:			
Bread	6	44*	38
Milk	0	0	2
Potato	51	48	33
With measurable residues			

* Residues of chlormequat assessed for first time and found in 41% of samples.

Analyses carried out by the Co-op indicated that stored products such as tree fruits, citrus, root and salad vegetables frequently contained residues (Table 11).

The most complete data set currently available in the UK remains that from PRC. (Table 12). A sampling of 53 vegetable crops and 22 fruit crops over the period 1991-2002 indicated that the % of samples with residues ranged from 100% to 0 in both vegetables and fruit. The frequency of occurrence was however higher in fruit. In addition in some crops e.g. lettuce, large numbers of individual compounds could be found. This data set allows changes in individual crops or of individual residues to be assessed with time (Table 13) for a limited number of

Table 10: Pesticide Residues Detected in Samples Analysed for the Pesticide Residue Committee January-June 2002

Crop/Commodity	No Samples Analysed (with Residues)	No Residues found (above MRL)	No fungicide found	No insecticide found
Apricot	16 (11)	20 (2)	16	4
Aubergine	22 (0)	-	-	-
Baked Beans	72 (0)	-	-	-
Bananas	64 (35)	45	45	-
Green Beans	36 (8)	9	6	3
Grapes	36 (21)	37	26	11
Leek	24 (0)	-	-	-
Orange	72 (51)	182	89	61
Peach/Nectarine	45 (28)	42	31	11
Pear	75 (47)	109 (1)	53	4
Potato	101 (55)	71 (6)	22	2
Pepper	36 (0)	-	-	-
Yam	32 (26)	26 (3)	26	-
Bread	48 (27)	28	3	-
Chocolate	48 (6)	6	-	6

Information on pesticide residues is also available from a limited number of supermarket websites. Residues detected by Co-op are summarised in Table 11.

Table 11: Pesticide Residues Found in Studies Carried out by Co-op

Date	No. of Tests Food X Individual Tests	% with Residue	No. of Residues	Problem Foods
Aug 2002	1680	28	11	Apricot, Apple, Lettuce
July 2002	2152	30	10	Apricot, Apple, Lemon, Orange
June 2002	1762	41	18	Apple, Orange
May 2002	2035	32	13	Grape, Beans, Apple
April 2002	1958	33	11	Celery, Tomato, Pear
March 2002	2144	61	33	Yam, Apple, Grape, Cucumber, Lettuce
Feb 2002	2984	56	30	Sprouts, Broccoli, Apple, Celery, Tomato
Jan 2002	2463	41	21	Apple, Orange, Lettuce, Pepper, Grapefruit

Table 12: Pesticide Residues Detected in UK Government Surveys Over the Period 1991-2002 (Data from PRC)

Crop	No. of Studies Conducted	Total No. of Samples Tested	No. Found with Residues (%)	No. of Pesticides Found	No. of MRL	No. of Pesticides on Co-op Restricted List
Vegetables and Herbs						
Chicory	1	12	12 (100)	1	0	0
Mooli	1	1	1 (100)	1	0	0
Celery	5	276	182 (66)	30	11	5
Carrot	6	369	235 (64)	12	3	1
Yam	2	58	36 (62)	2	3	1
Lettuce	12	803	463 (58)	37	30	5
Onion	3	146	70 (48)	1	0	0
Potato Crisps	2	180	86 (48)	5	0	0
Leaf Salad	1	24	11 (46)	6	0	0
Mange Tout	1	40	16 (40)	2	0	2
Pea in Pods	2	60	22 (37)	13	0	3
Potato (main crop)	19	1722	639 (37)	15	6	2
Radish	2	40	14 (35)	2	0	0
Potato (salad)	5	205	55 (27)	6	0	0
Tomato	6	359	84 (23)	26	1	4
Cucumber	4	215	48 (22)	12	2	4
Spinach	1	23	5 (22)	2	0	1
Parsnip	3	85	17 (20)	4	0	0
Cabbage (head)	1	72	11 (15)	6	0	1
Baby Veg	1	49	7 (14)	6	0	1
Cabbage (white)	1	24	3 (13)	1	0	0
Aubergine	2	42	5 (12)	3	0	0
Chilli	1	18	2 (11)	2	0	1
Mushroom	3	255	27 (11)	5	2	1
Pepper (sweet)	4	172	19 (11)	4	1	2
Artichoke	1	22	2 (19)	1	0	1

Contd...

Contd...						
Turnip	2	44	4 (9)	2	0	0
Cabbage (Chinese)	3	70	5 (7)	5	0	2
Broccoli	4	134	8 (6)	5	0	0
Beans (green)	3	70	4 (6)	4	1	3
Beans (broad)	1	16	1 (6)	1	0	1
Onion (spring)	2	59	3 (5)	2	0	0
Brussel Sprout	2	73	3 (4)	2	0	0
Pea (shelled)	2	72	3 (4)	2	0	0
Courgette	3	120	4 (3)	4	0	1
Calabrise	2	87	2 (2)	2	0	0
Cabbage (green)	3	71	1 (1)	1	0	0
Asparagus	1	26	0	0	0	0
Beetroot	1	19	0	0	0	0
Cabbage (red)	1	22	0	0	0	0
Cauliflower	2	45	0	0	0	0
Fennel	1	12	0	0	0	0
Garlic	2	48	0	0	0	0
Kohlrachi	1	6	0	0	0	0
Leek	2	84	0	0	0	0
Marrow	2	85	0	0	0	0
Okra	1	24	0	0	0	0
Olive	1	24	0	0	0	0
Spring Green	1	5	0	0	0	0
Squash	1	24	0	0	0	0
Swede	2	67	0	0	0	0
Sweet Corn	1	29	0	0	0	0
Water Cress	1	12	0	0	0	0
Clementine	1	16	16 (100)	11	0	1
Mandarin	1	34	34 (100)	13	1	1
Soft Citrus	2	162	162 (100)	24	3	3
Orange	4	303	287 (95)	30	6	4
Lemon	4	84	79 (94)	19	1	2
						Contd...

Contd...						
Apricot	1	24	19 (79)	8	2	2
Grapefruit	3	122	83 (68)	18	3	1
Raspberry	2	30	20 (67)	12	1	3
Strawberry	7	383	257 (67)	33	5	6
Banana	3	181	117 (65)	7	5	2
Currants	4	77	47 (61)	18	2	3
Gooseberry	1	25	15 (60)	6	1	3
Loganberry	2	12	6 (50)	2	1	1
Kiwi	4	106	52 (49)	8	0	3
Apple	2	396	173 (44)	25	0	3
Grape	8	382	167 (44)	46	8	6
Melon	1	72	30 (42)	16	2	4
Passion Fruit	1	19	7 (36)	6	2	2
Peach/Nectarine	6	316	109 (34)	27	6	4
Plum	3	138	40 (29)	8	1	1
Pineapple	2	49	11 (22)	1	1	1
Coconut	1	12	0	0	0	0

commodities. In potato the proportion of samples containing residues and the predominant residues found, growth regulators and fungicides, were relatively constant with time (Table 13). UK and imported samples had different profiles. Unlike the data available from USA (Baker *et al,* 2002) UK data does not appear to identify the types of production systems from which the produce sampled had been derived.

In 2001, PRC (2002) noted that "desirable safety margins" had been eroded by 0.25% for the 0.7% of samples exceeding MRLs, with the outcome, on a worst case scenario, being an upset stomach. Just over 70% of the produce tested, 4,003 samples in 2001 were free of pesticide residue. The equivalent figures for nine supermarket chains, as measured on behalf of Friends of the Earth (2001), were 27 to 71%. Food popular with children, ranging from infant food to pizzas, are subject to special consideration, as are foodstuffs that have particularly attracted attention. Thus it is not easy to establish trends as the range of products tested

Table 13: Pesticide Residues Found in Potatoes 1991-2001 by UK Government Surveys*

Year of Study	Source of Produce	% Samples with Residues	Principal Residues
2001	UK	33	Maleic hydrazide, chlorpropham
2000	UK	48	Maleic hydrazide, Chlorpropham
1999	UK	51	Maleic hydrazide, Chlorpropham
1997	UK	22	Maleic hydrazide, Chlorpropham
1995	UK	24	Maleic hydrazide, Chlorpropham
1994	UK	47	Maleic hydrazide, Chlorpropham
1993	UK	40	Maleic hydrazide, Tecnazene
1992	UK	42	Maleic hydrazide, Chlorpropham
1991	UK	45	Maleic hydrazide, Tecnazene
1995	Imported	11	Thiabendazole, Chlorpropham
1994	Imported	17	Chlorpropham, Thiabendazole
1993	Imported	9	Chlorpropham, Thiabendazole
1992	Imported	23	Maleic hydrazide, Tecnazene
1991	Imported	8	Tecnazene

* Data from FSA.

varies from one year to the next on all but the staples. For example, there was no report in 2001 on baked beans, bananas, biscuits, coffee, condiments, cooking oils, cooking sauces, curry spices, game, nuts, pickles and other preserves, sugar, tomato ketchup, nor on any alcoholic beverages.

As indicated above information on crops containing residues is available from both government and some supermarket sources. Commodities identified both by government (2001) and Co-op (Sept 2001-Aug 2002) as often containing residues are summarised in Table 14. This listing is restricted to commodities that can be produced in UK (although not all those identified here as having residues were produced in UK).

These listings suggest that changing pesticide use in top and soft fruit, leafy vegetables and potato would have the largest impact on the general exposure to pesticides.

Table 14: Commodities Identified as Commonly Containing Pesticide Residues*

Source			
Co-op	**No. of Samples Found with Residues**	**Pesticide Residue Committee**	**No. of Samples Found with Residues**
Lettuce*	6	Apple	68
Mushroom	1	Mushroom	14
Apple	18	Courgette	03
Beans	1	Potato	80
Strawberry	2	Strawberry	115
Celery	2	Celery	42
Tomato	3	Tomato	26
Cucumber	1		
Brussels Sprouts	2		
Broccoli	2		
Capsicum	4		
Pear	4		
Leek	1		

* Residues Above MRL Detected.

The status of pesticides in foods consumed in New Zealand has been reviewed by Cressey *et al* (2000). They found that of the 460 samples which they screened for 90 pesticides that 59% contained detectable residues. A similar survey in 1990 found 56% of samples to contain residues. In their 1997 survey the number of distinct residues found fell to 20 substances compared to the 30 found in the earlier studies. The increased detection of residues in 1997 was a function of improved analytical technology. They commented that of the 272 samples with residues found 160 would not have been detected in the earlier study with the methods used at that time. The New Zealand Study detected 5 (of 23 tested for) OC residues, including DDT endosulfan and dicofol, 5 (of 33 tested for) OP residues including chloropyriphos and dimethoate and 7 (of 18 tested for) fungicides, including iprodione, chlorothalonil, vinclozolin and dithiocarbonates. This would suggest that the current status of residues in fresh produce in New Zealand, a major source of produce on the UK market is similar to that in UK.

Maximum Residue Levels *Codex Alimentarius*, are usually measured in mg kg^{-1} or parts per million, but in some cases are based on current Limits of Detection (LOD, hence LOD-MRL). There is also a programme of harmonisation within the European Union (EU-MRL). PRC's reports are directed at an informed public. There is, however, often failure to successfully to put across the message that MRLs are not safety limits i.e. they can be exceeded without implying a risk to health, but that their primary role is to demonstrate the pursuit of Good Agricultural Practice (GAP), thus to facilitate international trade.

Adherence to Good Agricultural Practice (GAP) requires a different approach to that associated with food safety. Proof of failure to comply with GAP needs residue tests based on whole, unpeeled, uncooked produce to demonstrate exceedance of MRL. Sample preparation to estimate SMTR for dietary exposure in the USA (Lamont, 2002) requires produce to be cleaned, peeled and, if necessary, cored so as to produce the material that is consumed; this is also not cooked because a processing factor is included in the dietary exposure calculation.

The calculation of ADI, expressed in mg active ingredient per kg body weight, is based on a complex model of risk factors coupled to the "No Observed Adverse Effect Level" (NOAEL) in animal experiments, but multiplied up by an uncertainty factor, usually one hundredfold. It is not clear to what extent ADI as a risk assessment takes into account residue losses during processing, and residue gains associated with, for example, the use of spices and the treatment of food premises (Singh & Singh, 1990).

Variation in the residues found in fruit and vegetables has recently been reviewed by Hill and Reynolds (2002). They assessed pesticide residues in individual units taken from large samples of apples, bananas, celery, kiwi, oranges, peaches, peas, plums, potatoes and tomatoes drawn from 17 countries. Analytical variance contributed less than 11% to total variance. They found that the primary determinant of variability was the quantity of pesticide which initially arrives (by deposition or translocation) at the time of treatment. Reducing variability will therefore depend when improving the uniformity of the application of pesticides. This would suggest that variability is unlikely to reduce, in a major way, in the foreseeable future. Apparent residues of some materials can be influenced by binding as discussed by Skidmore *et al* (2002).

3.9 Pesticide Residues in Processed Food and Beverages

A combination of washing, blanching, and cooking removes the majority of pesticide residues from vegetables and fruit. For example, 83% of benomyl residue was removed by washing tomatoes, and washed and peeled tomatoes lost 99% of their carbaryl and malathion residues (Elkins, 1989). Given current concerns about maneb residues, Elkins' findings for processing leafy greens and spinach are of interest, with 74 to 94% of residues lost in washing, albeit a more rigorous washing process than might be associated with domestic cookery preparation. This may explain why normal washing of plant products has also been found to have no effect (Dejonkheere *et al.*, 1996). The same study demonstrated the overriding value of peeling for a much wider range of pesticides and commodities, with steaming and cooking varying in impact; however maneb appeared to form one of the most dislodgeable fungicide residues, 64% being lost from washed lettuce as opposed, for example, to 43% for vinclozolin. Elkins (1989) demonstrated a further 83-92% loss during blanching and canning. The concentration of ethylene-thiourea (ETU), the main degradation product of maneb, increased during canning up to 1.1 ppm or less.

Tomatoes processed into purée and ketchup lost 98% of benomyl residue, and the most concentrated tomato paste had levels of pesticide below that of the initial product (Elkins 1989). Oranges processed into juice, concentrate and orange oil respectively lost 98%, 90% and 23% of their original benomyl residue.

Drying plant material results in a concentration of residues, but this varies between pesticides and there are some losses during the process. A good example is provided in the prune industry in France (Reulet *et al.* 1998) with the increase in concentration of dithiocarbamates being very low, carbendazim residues in prunes being twice those of plums, and iprodione higher again (2.6 times), concentration of plums to prunes being on average 3.2. Possibly of more relevance to British consumption is the potential for concentration of pesticide residues in dried herbs and spices. No reference literature referring to this could be found.

A recent study of wine corks revealed the almost ubiquitous presence of organochlorine residues derived from aldrin, chlordane, DDT, dieldrin, endosulfan, endrin and hexachlorocyclohexane (HCH), but in concentrations at the ng g^{-1} (i.e. p.p.b.) level (Strandberg & Hites 2001).

3.10 Multiple Dosage

The cocktail effect, has been given active consideration recently by the Working Group for the Risk Assessment of Mixtures of Pesticides (WiGRAMP) of the Committee on Toxicity of Chemicals in Food, Consumer Products and the Environment. The conclusions were that the health hazard associated with mixtures was almost certainly small with those often considered more at risk than the majority of the population no more vulnerable than general, but that data are lacking and some interactions are not predictable.

The need for additional precautions to minimise the effects of pesticides being combined within the diet has recently been reviewed by Carpy *et al.* (2000), based on a 14 year literature review. They conclude that, despite there being a large body of knowledge on risk assessment for exposure to chemicals in mixture, there is no single methodology available, i.e. each case must be considered using professional judgement. Exposure to a mixture of compounds does not generally cause effects greater than those of their most active component, provided one is dealing with low concentration levels such as ADI and the reference dose level (RfD), which are well below the NOAELs. The use of an additional safety factor to reduce such levels is not supported.

This, of course, is in contrast to the findings of Arnold *et al.* (1996), a former researcher at Tulane University, USA. His laboratory produced results falsified to imply that some organochlorine insecticides and PCBs, which have weak oestrogenic activity when tested alone, were up to 1,600-fold more potent in mimicking oestrogen when tested in combination. Although the original paper was subsequently withdrawn from *Science* (McLachlan 1997), and Dr. Arnold was noted as being engaged in scientific misconduct (Office of Research Integrity, 2001), this study nevertheless led to passage of the Food Quality Protection Act (FQPA) in the USA in 1996, requiring the EPA to develop a programme to screen the endocrine disrupter capacity of many chemicals.

Although the cocktail effect may also be influenced by the processes of food blending and consumption, some crops (Friends of the Earth, 2002) frequently contain multiple residues. Baker *et al.* (2002) noted that conventionally grown products had more multiple pesticide residues than did organic produce.

The significance of a pesticide can also be influenced by the environment in which it is found. For example Jigang and Laird (2003) found that when the OP insecticide chlorpyrifos was mixed with chlorinated tap water that it was converted, by the HOCl to chlorpyrifos oxon a much more toxic cholinesterase.

A study of pesticide residues in food derived from conventional, IPM or organic systems showed that pesticide use influenced the presence and profile of residues. The study showed that organic foods had around one third of the frequency of residues found in conventional foods and half that found in IPM derived food. Organic foods had fewer multiple residues. The concentration of residues was lower in organic samples. IPM samples tended to have lower residues than normal samples.

3.11 The Consequences of Reducing Pesticide Use

The role of pesticide use in driving the design of current food production systems means that changes in pesticide use would impact on a range of issues from fertiliser use strategy to food quality and patterns of labour deployment. An example of the complex effects can be seen in the study of the advantages and disadvantages of using fungicides by Schneider and Dickert (1994). Reducing pesticides application is likely to influence the appearance of fruit and vegetable products and may change their chemical and biochemical properties unless a range of other changes are instituted in parallel with a change in pesticide use. Not all consequences of reducing pesticide use would be positive.

The complex background into which changes in pesticide use would need to be fitted is exemplified by considering herbicide use. A review of herbicide use and invention (Parry 1989) began with the question "why use herbicides?". The answers given were:

a) to reduce competition for water and light from other species (weeds);

b) to improve crop management especially harvesting and to raise the quality of the final harvested product; and

c) to reduce the risk of cross infection from fungi and insects to the crop.

Outwith the biological sphere, to reduce the costs of production and especially the costs of manpower, could have been added. Similar and parallel answers could be given to questions about fungicides and insecticides. The purpose of the use of pesticides is to increase resource allocation to the crop species, to aid crop management, to improve the economic efficiency of resource use and to simplify management. On the basis that modern systems of agricultural production have been developed with the intention of using pesticides then a reduction or the removal of pesticides from the system will have major effects. As detailed below (3.12) ending pesticide use would require the development of systems where pesticide inputs are replaced by non-chemical inputs which serve the same purpose or alternatively that the system is wholly redesigned so that controls are effected in other ways; this is of course the objective of organic or ecological agriculture.

The development of reduced pesticide systems is conceptually more difficult than non-pesticide systems. They must result from a system in which applications of pesticides are made less often, at reduced rates or deliberately through using materials less likely to result in residues or through all of these. This approach risks competition from weeds (or damage from pests or diseases) not being sufficiently controlled, with adverse effects on yield and quality. It also requires a series of decisions on whether or not to spray and how much to spray. This is likely to require an increase in manpower to carry out the monitoring which must inform such decisions. Reducing the amounts of pesticides applied thus brings with it increases in costs and risks to the volume and quality of supply. On the basis that pesticide residue levels in foods are at least in some part a function of pesticides applied it would have the benefit of reducing residues. Much of this review is focussed on the options available for reducing pesticide use in current production systems. In the short term this is likely to be the easiest way ahead. Variation in the residues found in crops (Tables 1-4) and the impact of production system on the levels of residues (Baker *et al*, 2002) shows that this approach can have an impact on residues.

3.12 The Consequences of Ending Pesticide Use

Many of the pesticides used in the past have left significant residues in the soil. Ending pesticide use in the UK would not therefore remove all residues. The study of Baker *et al* (2002) identified that foods produced by organic agriculture were not residue free. The proportion of organic samples containing pesticide residues at 23% was significantly lower than the 73% of samples from conventional agriculture. Organic crops with unusually high occurrences of residues included peach, broccoli, celery and spinach. Many of the residues were those of older organochorine pesticides. Excluding these reduced the proportions of organic and conventional crops with residues to 13 and 71% respectively. The highest levels of individual pesticides were for chlorpropham in potato (1.6 ppm), permethrin in spinach (0.49 ppm), methamidophos in capsicum (0.68 ppm), formetanate in orange (0.3 ppm), vinclozolin in grape (0.2 ppm), methyl in strawberry (0.19 ppm) and iprodione in grape (0.14 ppm). Other levels were below 100 ppb. Ending pesticide use in UK could therefore benefit organic agriculture in the short term and perhaps GM crops in the longer term. Many first generation GM crops have however been engineered to give resistance to specific herbicides with broad-spectrum activity. The 2001 of the Pesticides Residues Committee for 2000 identified the presence of glyphosate residues in 7% of bread samples. Wider use of such broad-spectrum materials would be expected to increase the frequency with such residues are found. The use of glyphosate as a desiccant is probably the main reason for this increase. The use of strobulurin fungicides, which keep the cereal canopy green for longer, may have intensified desiccant use. Increasing the number of crops resistant to glyphosate may accentuate the current trend. Discussing the impact of a zero base line is important to identifying the scale of reduction, which might be achievable. Reduced and zero pesticide systems need to be assessed for individual crops in the field and post harvest. The extent to which major field crops could be produced in UK without pesticides, at a quality acceptable to consumers and at a price sustainable in a world market is a very real issue. Against an overall background of minimising residues in all products so that levels which exceed MRLs become increasingly rare there may be a continuing separate market placing for both non and low pesticide products (currently organic and others). This gives rise to two distinct groups of questions, i.e. how can the productivity and quality of organic crops be increased without resorting to chemical control and for other systems of production how can the amount of pesticide finding its way into final product be minimised while maintaining current high standards of crop protection? These issues are the main questions for Sections 5 and 6 of this report.

3.13 Conclusions on Pesticide Residues

a) Losses of crop due to fungal diseases, pests and weeds remain significant considerations for those involved in crop production in UK and elsewhere.

b) Variation in the % of crop tested containing pesticide residues and in the number of residues found suggests that there exists great scope to move both the average residue level and the mean % of samples with residues to current minimum levels. The establishment of a best practice approach would be helpful.

c) Much current variation has been identified as being at least in part a consequence of application practice. This suggests scope for the agricultural industry to reduce current pesticide levels. This would be aided to produce if the need to do so became a priority and a design feature of agricultural systems.

d) A high proportion of current residues are represented by materials applied postharvest, both growth regulators and fungicides. Identifying the scope to reduce the use of such materials must be an important component of a strategy.

e) Of the residues found, which did not relate to post-harvest use fungicides were the most common. The development of means of reducing plant diseases without resorting to prophylactic applications of fungicides will be important to reducing this major category of use. The successful use of fungicides such as the strobilurins to protect the leaf canopy has resulted in increased use of desiccants, which are now being found as residues in products such as bread.

f) The availability of residue data for products produced by different agricultural systems e.g. ICM, organic, conventional, as has used in some USA studies e.g. Baker *et al* (2002) is essential to assessing the impact of changes in agricultural husbandry on the generation of residues.

g) Those involved in developing pesticide application strategies need to be more aware of the links of practice to residues in foods.

h) Data on pesticide residue levels for the following would be of value:

- Freshly harvested fruits, vegetables and herbs

- Grain from Intervention storage and other situations with post-harvest pesticide treatment
- Potatoes subject to long term storage
- Refined products such as sugar
- Crop food products prepared for the table
- Beers, wines, cider, and fruit juices
- Milk, including the extent to which residues are concentrated or lost in milk products
- Cooking oils, including the extent to which residues are reduced by heat
- Seasoning and preservative spices
- Organic products.

i) The establishment of priorities must be informed by what is known of the residue levels through the food chain. The toxicological pack for all new pesticides includes an examination of degradation and metabolism pathways, but this needs to be supported by evidence of what happens in practice.

(Atkinson D, SAC – Edinburgh, The Scottish Agricultural College (SAC), United Kingdom,

F Burnett, Researcher (Plant Pathology), Crop & Soil Systems Research Group King's Buildings, Edinburgh,

G N Foster, can be reached at g.foster@au.sac.ac.uk,

A Litterick, Environmental Consultant, Ferguson Building, Craibstone Estate, Aberdeen,

M Mullay, Web Editor, SAC, West Mains Road, Edinburgh,

C A Watson, SAC – Aberdeen, The Scottish Agricultural College (SAC), United Kingdom.)

6

The Rotten Apples
Experiences of Agrochemicals and Food Safety in India and Other Countries in the Developing World

Amrita Chakraborty

The ordinary readers, the common man comprising mothers, fathers and others, must know that apples, peaches and lettuce and many other 'good food' that they consume and give their children have traces of cyhexatin, fenhexamid, glyphosate, indoxacarb and many other residues of pesticides and agrochemicals that they are not even aware of. It may not be far fetched to say that today's apples are somewhat different from what Adam & Eve had had! Along with God given nutrients they contain residues of acephate, cyhexatin, indoxacarb etc. And we have the right to know more explicitly about the pesticides and agrochemical residues that 'good food' like fruits and vegetables and milk have.

A Joint FAO/WHO Meeting on Pesticide Residues (JMPR) held at World Health Organization headquarters, Geneva, from 20 to 29 September 2005 brought together the FAO Panel of Experts on Pesticide Residues in Food and the Environment and the WHO Core Assessment Group[1]. JMPR

plays an important role in the improvement of food safety on a global basis, by laying the scientific foundation for the development of international and national food standards. And a wider percolation of their findings amongst the common man is both important and urgent.

Introduction

Food, life supporting substance, that everybody requires from birth till death. A basket of food, that ubiquitous pot-pourri conjures up so many thoughts in our minds. It is essential for life, tasty, nutritious, safe, aromatic, romantic, and replete with mother's love and many wonderful connotations. Add to this, what scientists tell us repeatedly about fruits and vegetables, that they are ranked high in a hierarchy of good food. Good food is apart from junk food. What, then, is good food? Is it nutritious food? Or is it food that is safe to consume? Surely food can not be 'good' if it threatens life, whether in the short run or even in the long run. Food is essentially life supporting and even life generating. Then, in the name of nutrition, is it acceptable that we keep taking food that may be life threatening? These questions and many more arise when we read and hear about the extent of pesticides in fruits and vegetables in developing countries.

Most developing countries are constrained by a large population, hence, many mouths to feed and they use large quantities of pesticides and other agrochemicals to increase productivity from land. The pesticides and other agrochemicals not only enter into the current harvest but leach into the soil and percolate into succeeding crops and ground water. Thus, making crops pesticides contaminated and unsafe. If the contamination is higher than acceptable limits, then crops can become life threatening than life supporting. Then how can an apple a day keeps the doctor away? Apples are among the so-called "Dirty Dozen" of fruits and vegetables sold in the United States.

Every year the world uses 3 million tons of chemical pesticides. In the US, 62% of crops are treated with some form of pesticide, but it is estimated that less than 1% of the applied pesticides actually reach the target organism[2]. The remaining ends up in the air, water and soil and most importantly, on (and in) our fruits and vegetables. So how safe is our food? The 'good food' comprises fruits and vegetables, leave alone processed food which has added chemical preservatives, colours etc.

However, the countries in the developed world, especially the US, do consider food safety as an important aspect of good food and implements food safety very stringently. It has assigned clear responsibilities to two nodal agencies—the USEPA and the US Food and Drug Administration (USFDA). The former is the standard-setting agency. It is entrusted with registering pesticides for use. Before registration, it establishes ADI, Acceptable Daily Intake—what it calls Chronic Reference Dose (CRfD)[3]—for a pesticide and sets MRLs (maximum residue limits or tolerance limits) for residues on food commodities. What is pertinent for us in India is that USEPA & USFDA tries to make sure that exposure in a food commodity is below ADI. The EU process is quite similar to what the US does.

However the picture is very different in developing countries. China which has emerged in the 1990s as a large low-cost producer and exporter of food products such as vegetables, apples, seafood, and poultry saw exports slowing when shipments of vegetables, poultry, and shrimp were rejected for failing to meet stringent standards in Japan, Europe, and other countries, revealing a gap between Chinese and international food safety standards. In 2002 Chinese frozen spinach in Japan was found to have high levels of a pesticide.

Problems with food contamination within China have made food safety a top concern of Chinese consumers as well. Many of China's food safety problems can be traced back to the farm level. Farmers rely on heavy use of chemicals to coax production out of intensively cultivated soils and deal with pest pressures, a practice that contributes to food safety problems. China has one of the world's highest rates of chemical fertilizer use per hectare, and Chinese farmers use many highly toxic pesticides, including some that are banned in the United States. Some farmers have little understanding of correct chemical use; for example, they may fail to wait the prescribed number of days between the last application of a pesticide and harvest, resulting in excessive residues in the harvested product[4]. However, the Chinese government is tackling this problem with several sincere efforts.

The Government has responded by trying to build a food safety system for exports that will establish China's international reputation for producing safe food. China has also been raising domestic food safety standards and implementing inspection and testing systems for consumer products and agricultural commodities. What is important for Indian readers to know is the Chinese government is sensitive enough to this problem because in 2005, officials announced plans to update a 1995 law covering consumer food products. In 2006, the Chinese legislature adopted a law that establishes a national framework for building a system that ensures the safety and monitoring of agricultural products. Local governments have also been active in promoting safer food.

The Indian Scenario

What about the pesticides exposure in fruits, vegetables and milk in India? How safe are the Indian fruits and vegetables? Are they really 'good food' in terms of food safety? Specially, in the light of the recent brou-ha-ha over pesticides in soft drinks, many people are interested to know what is the story of pesticide residues in fresh foods in India?

In global practices, the agency registering the pesticide establishes ADI, sets MRLs and then ensures cumulative exposure is within the safety levels. In stark contrast in India, a pesticide is registered without any of these mandatory safety regulations.

In fact, there is no legislative provision to link pesticide registration to setting MRLs. IA (Insecticide Act, 1968) mandates registration, but PFA, the Prevention of Food Adulteration Act, 1954 mandates MRLs. Such legislative blindness has ensured that, of the 180 pesticides currently registered, MRLs have been set only for 71[5]. In other words, more than 60 percent of pesticides currently registered have no MRLs.

Take a look at the following 'good food'(?) and decide whether the statistics makes us feel uneasy or not.

1. *The case of Indian sugar cane:* The CIB (Central Insecticide Board) recommends 13 pesticides to be used. However, under PFA, MRLs for

only 2 of the recommended pesticides have been established.[6] Does this mean that Indian sugar cane is safe (and good) when it meets the MRLs for the 2 recommended pesticides? Specially, against the backdrop of not having MRLs for 10 or more pesticides!!

2. *The case of Indian rice:* 56 percent of recommended pesticides have no MRLs; if 44 percent of pesticides don't even have MRLs, then how do we know the safety of rice that most of us take in large quantities everyday!
3. *The case of Indian wheat:* 43 percent have no MRLs.
4. *The case of Indian mango:* 44 percent have no MRLs. It is the king of fruits and the Indians love it. People have mangoes from breakfast to after dinner and imagine we don't have MRLs for 56 percent of the pesticides that are going into them.
5. *The case of Indian coffee:* 80 percent of recommended pesticides have no MRLs[7].

Contamination of food and water that most people eat and drink is quite dangerously high in India. In general, fruits and vegetables and milk are India's most contaminated food. Yet, renowned research organizations to catch eye-ball attention would rather take on Coke & Pepsi and not highlight the poison in the general food basket like Indian sugar cane, Indian rice, Indian mango etc (consumed by most people in much larger quantities than fun food like the bubblies)!!!!!!! Which is a greater poison, a few Cokes or Pepsis now and then or generous helpful of apples and peaches and lettuce, generously spiked with 'recommended' pesticides! But surely gunning for pesticides in products, a poor farmer's field may not fetch the necessary eye-ball attention as one may get attacking MNCs!!

Even ancient healing methodologies like homeopathy use poisons like arsenic etc as medicines. Why and how does arsenic heal when it is a known poison and when consumed in higher quantities is debilitating. The right dose differentiates a poison and a remedy. Hence, modern, educated consumers, like us, should be aware of ADIs and ARfDs & MRLs, the right doses that we should not overstep.

And food regulators in our country should determine and raise our awareness about what that right dose of pesticides is in the various fruits and vegetables and milk that really forms our daily diet.

The Important Facts

The first chemical pesticides to see worldwide use were arsenic-base compounds such as Paris Green and London Purple, referring to their derivation from paint pigments[8]. The right dose differentiates a poison and a remedy like homeopathy's use of poisons like arsenic has shown us. The right doses are estimated through MRLs, ADI and ARfD. Hence, ADI and ARfD are crucial tools to manage health risk through food. For them to be consistently effective, they need to be (a) constantly updated as science improves; and (b) calculated on the basis of the latest, most credible data.[9] Though ADIs and ARfDs, in India, are ignored unpardonably, their gravity is immense in developed countries, in FAO, WHO etc.

A Joint FAO/WHO Meeting on Pesticide Residues (JMPR) held at World Health Organization (WHO) headquarters, Geneva, Switzerland, from 20 to 29 September 2005 brought together the FAO Panel of Experts on Pesticide Residues in Food and the Environment and the WHO Core Assessment Group.[10] JMPR plays an important role in the improvement of food safety on a global basis, by laying the scientific foundation for the development of international and national food standards. It has been accepted that the work of the Meeting is seen as an integral part in the safe use of pesticides to ensure food security and for overall sustainable development.[11] The FAO and WHO are some of the highest and most prestigious bodies who study (undertaken jointly by experts) to evaluate possible hazards to humans arising from the occurrence of residues of pesticides in foods.[12]

The Joint FAO/WHO Meeting on Pesticide Residues (JMPR) used the spreadsheet to obtain maximum residue level estimates for several of the compounds considered. The findings were compared with the estimates made independently by evaluators. Generally, the recommendations of the spreadsheet compare favourably with those determined by the evaluator. The comparison is summarized in the following Table 1:

Table 1: Comparison of Some MRLs as Determined by JMPR and by Spreadsheet

Commodity	No.	High Residue (mg/kg)	Median Residue (mg/kg)	MRL JMPR (mg/kg)	Spreadsheet MRL		
					Source	Estimate (mg/kg)	MRL (JMPR Rounded) (mg/kg)
CYHEXATIN							
Orange	20	0.10	0.05	0.2	99LN	0.22	0.3
Apple	48	0.16	0.03	0.2	$\mu+3s$	0.16	0.2
Grapes	36	0.19	0.08	0.3	$\mu+3s$	0.22	0.3
FENHEXAMID							
Cherries	20	4.7	1.35	7	LN99	5.38	7
Peaches	12	5.9	3.85	10	$\mu+3s$	8.79	10
Plums	27	0.79	0.31	1	$\mu+3s$	0.91	1
Grapes	11	11	4.3	15	$\mu+3s$	13.42	15
Strawberry	6	5.9	3.3	10	$\mu+3s$[1]	10.29	15
Bushberry	16	2.9	1.65	5	LN99	3.68	5
Caneberry	13	11	2.0	15	UPLMed95	14.21	15
Kiwi	9	11	6.3	15	LN99	14.88	15
Cucumber	16	0.65	0.185	1	$\mu+3s$	0.60	0.7
Tomato	17	0.93	0.04	2	LN95	1.17	2
Pepper	18	1.5	0.71	2	LN99	1.60	2
Lettuce	8	19	11.5	30	$\mu+3s$[2]	29.78	30
Almonds	5	0.02	0.02	0.02	$\mu+3s$	0.02	0.02
GLYPHOSATE							
Beans, dry	19	1.8	0.17	2	Ln99	2.19	3
Peas, dry	11	2.1	0.50	5	UPLMed95	3.81	5
Soya beans	36	17	1.85	20	LN95	16.11	20
Maize	21	3.0	0.05	5	$\mu+3s$	2.16	3
Cereal grains (ex maize and rice)	84	20	3.85	30	LN95	30.92	40
Cotton seed	23	28	5.00	40	$\mu+3s$[3]	33.97	40
Rape	35	12	0.96	20	LN95	17.46	20
Sunflower	8	5.6	0.40	7	$\mu+3s$[4]	9.11	10

Contd...

Contd...

Alfalfa hay (fodder)	23	341	189	500	LN99	568.15/.89	700
Grass hay	13	259	187	500	μ+3s	409/0.88	500
Bean fodder	10	93	22.5	200	UPLMed95	179.46/0.90	200
Pea fodder	10	320	102	500	LN99	710.27/0.88	800
Barley straw	27	160	47	400	LN95	356.81/0.88	800
Maize fodder	20	92	20.5	150	UPLMed95	126.85/0.83	160
Oat straw	11	27	64	100	LN99	145.81/0.90	160
Sorghum fodder	10	33	18.5	50	μ+3s[5]	51.35/0.89	60
Wheat straw	29	198	47	300	LN95	327.72/0.88	400
INDOXACARB							
Apple	14	0.30	0.21	0.5	LN99	0.42	0.5
Pear	6	0.11	0.06	0.3	LN99	0.13	0.2
Peach	9	0.18	0.11	0.3	LN99	0.28	0.3
Grapes	16	1.5	0.30	2	LN99	2.5	3
Cabbage	8	2.7	0.44	3	UPLMed95	3.95	5
Broccoli	8	0.14	0.06	0.2	LN99	0.24	0.3
Cauliflower	19	0.14	0.02	0.2	μ+3s	0.13	0.2
Cucumber	13	0.10	0.02	0.2	μ+3s	0.10	0.1
Melons	18	0.30	0.11	0.5	μ+3s	0.08	0.1
Tomato	8	0.30	0.11	0.5	LN99	0.57	0.7
Pepper	30	0.21	0.04	0.3	μ+3s	0.19	0.2
Sweet corn	12	0.01	0.01	0.02*	μ+3s	0.01	0.02
Head lettuce	9	4.3	2.8	7	μ+3s	6.27	7
Leaf Lettuce	9	8.4	6.6	15	LN99	14.27	15
Pulses	7	0.13	0.02	0.2	μ+3s	0.16	0.2
Soya bean	20	0.45	0.03	0.5	μ+3s	0.50	0.5
Potato	17	0.0085	0.003	0.02	μ+3s	0.01	0.01
Peanuts	13	0.003	0.003	0.02*	μ+3s	0.003	0.003
Cotton seed	7	0.92	0.36	1	LN99	2.25	3
Peaunt hay	12	45	16	50	LN99	90.74	100
Alfalfa hay	43	43	17	60	LN95	42.81	50

Contd...

Contd...							
Maize fodder	5	15	7.8	25	LN99	23.23	25
Cotton gintrash	7	11	8.0	20	LN99	16.19	20
METHOPRENE							
Cereal grains	12	8.1	4.85	10	LN99	13.83	15
NOVALURON							
Pome fruit	37	1.8	0.65	3	LN95	1.76	2
Soya	11	0.01	0.01	0.01	μ+3s	0.01	0.01
Cotton seed	16	0.40	0.07	μ+3s	0.48	0.5	
TERBUFOS							
Banana	21	0.03	0.01	0.05	μ+3s	0.03	0.03
Sugar beet tops	26	0.82 (3.56 dry)	0.05	5 (dry)	μ+3s	0.54 (2.35 dry)	3

[1] Rejected spreadsheet finding of a log normal situation (29 mg/kg.)

[2] Rejected spreadsheet finding of a log normal situation (72 mg/kg.)

[3] Rejected spreadsheet finding of a log normal situation (96 mg/kg.)

[4] Rejected spreadsheet finding of a log normal situation (3.59 mg/kg.)

[5] Rejected spreadsheet finding of a log normal situation (97 mg/kg.)

Source: Pesticide residues in food – 2005, http://www.fao.org/ag/AGP/AGPP/Pesticid/JMPR/DOWNLOAD/2005_rep/report2005jmpr.pdf

The maze of toxicological and scientific data shown in Table 1, Table 2 and Table 3 may at first glance, frighten the common man. But it has been gleaned from the highly credible and respected source, '*Pesticide residues in food – 2005*', a FAO report to serve as eye-openers for the common man in developing countries, unaware of the widespread contamination of pesticide residues in fruits and vegetables.

When prestigious bodies like FAO and WHO study (jointly by experts) and evaluate possible hazards to humans arising from the occurrence of residues of pesticides in foods and painstakingly tabulate the limits of daily intakes of rampantly used pesticides even region and ethnicity wise, it's of great interest to us to know whether our government adheres to such food safety standards and we receive the necessary benefit from such food safety standards. For instance, the following tables on International Estimated Daily Intakes of Pesticide Residues, given for two pesticides, give us the extent of pesticide an African or Latin American or European or a person from Mid-East or Far-East takes in each category of food.

Table 2: International Estimated Daily Intakes of Pesticide Residues

IEDI of Acephate

The following table give details of the international estimated daily intakes of the pesticides evaluated by the meeting for the five GEMS/Food diets and show the ratios of the estimated intakes to the corresponding ADIs.

(*) at or about the LOQ.

The ranges of the ratios of intake: ADI for all the compounds evaluated are tabulated in Section 3.

Diet: g/person per day; intake: daily intake: µg/person.

Acephate (95)		International Estimated Daily Intake (IEDI)									ADI = 0.0.03mg/kg bw	
		Diets:g/person/day										
Codex Code	commodity	STMR or STMR-p mg/kg	Mid-East diet	Mid-East intake	Far-East diet	Far-East intake	African diet	African intake	Latin American diet	Latin American intake	European diet	European intake
FP 0226	Apple	0.81	7.5	6.1	4.7	3.8	0.3	0.2	5.5	4.5	40.0	32.4
JF 0226	Apple juice	0.81	4.5	3.6	0	0.0	0	0.0	0.3	0.2	3.8	3.12
VS 0620	Artichoke globe	1.55	2.3	3.6	0.0	0.0	0.0	0.0	0.0	0.0	5.5	8.5
VP 0061	Beans except broad bean & soya bean (green pods & immature seeds)	1.35	3.9	5.3	0.9	1.2	0.0	0.0	4.4	5.9	13.2	17.8
VB 0400	Broceoli	0.16	0.5	0.1	0.1	0.2	0.0	0.0	1.1	0.2	2.7	0.4
VB 0401	Broceoli, Chinese	0.16	ND	-	ND	-	ND	-	ND	-	ND	-
VB 0404	Cauliflower	0.16	1.3	0.2	1.5	0.2	0	0.0	0.3	0.0	1.3	2.1
MO 0105	Edible offal (mammalian)	0.022	4.2	0.1	1.4	0.0	2.8	0.1	6.1	0.1	12.4	0.3
PE 0112	Eggs	0	14.6	0.0	13.1	0.0	3.7	0.0	11.9	0.0	37.6	0.0
FC 0003	Mandarins (incl. Mandarin-like hybrids)	1.15	8.8	10.1	0.2	0.2	0.0	0.0	6.3	7.2	6.0	6.9
MM 0095	Meat from mammals other than marine mammals: 20% as fat	0.022	7.4	0.2	6.6	0.1	4.8	0.1	9.4	0.2	31.1	0.7

Contd...

Contd...

MM 0095	Meat from mammals other than marine mammals: 80% as muscle	0.022	29.6	0.7	26.2	0.6	19.0	0.4	37.6	0.8	124.4	2.7
ML 0106	Milks	0.011	116.9	1.3	32.1	0.4	41.8	0.5	160.1	1.8	289.3	3.2
	Peaches & nectarines	1.35	2.5	3.4	0.5	0.7	0.0	0.0	0.8	1.1	12.5	16.9
FP 0230	Pear	0.81	3.3	2.7	2.8	2.3	0.0	0.0	1.0	0.8	11.3	9.2
VO 005	Peppers	1.9	3.4	6.5	2.1	4.0	5.4	10.3	2.4	4.6	10.4	19.8
PM 0110	Poultry meat: 10% as fat	0	3.1	0.0	1.3	0.0	0.6	0.0	2.5	0.0	5.3	0.0
PM 0110	Poultry, meat: 90% as muscle	0	27.9	0.0	11.9	0.0	5.0	0.0	22.8	0.0	47.7	0.0
VD 0541	Soya bean (dry)	0.055	4.5	0.2	2.0	0.1	0.5	0.0	0.0	0.0	0.0	0.0
OR 0541	Soya bean oil, refined	0.023	1.3	0.0	1.7	0.0	3.0	0.1	14.5	0.3	4.3	0.1
	Total intake (μg/person)=			44.0		13.9		11.7		27.9		124.1
	Bodyweight per region (kg bw)=			60		55		60		60		60
	ADI (μg/person)=			1800		1650		1800		1800		1800
	% ADI=			2.4%		0.8%		0.7%		1.6%		6.9%
	Rounded % ADI=			2%		1%		1%		2%		7%

Table 3: IEDI of Cholorpropham

Chlorpropham (201)		International Estimated Daily Intake (IEDI)										ADI = 0.0.05 mg/kg bw
		Diets:g/person/day										
Codex Code	Commodity	STMR or STMR-p mg/kg	Mid-East diet	Mid-East intake	Far-East diet	Far-East intake	African diet	African intake	Latin American diet	Latin American intake	European diet	European intake
VR 0589	Potato	3.60	59.0	212.4	19.2	69.1	2.6	74.2	40.8	40.8	240.8	866.9
MM 0812	Cattle meat: 20% as fat	0.004	2.9	0.0	0.5	0.0	2.1	0.0	6.0	0.0	12.7	0.1
MM 0812	Cattle meat: 80% as muscle	0.004	11.7	0.0	2.2	0.0	8.3	0.0	24.0	0.1	50.6	0.2
ML 0812	Cattle milk	0.0003	79.5	0.0	23.2	0.0	35.8	0.0	159.3	0.0	287.0	0.1
	Total intake (μg/person) =			212.5		69.1		74.2		147.0		867.2
	Bodyweight per region (kg bw) =			60		55		60		60		60
	ADI (μg/person) =			3000		2750		3000		3000		3000
	% ADI =			7.1		2.5		2.5		4.9		28.9
	Rounded % ADI =			7		3		2		5		30

Notes:

1. GEMS/Food Global Environment Monitoring System – Food Contamination Monitoring and Assessment Programme
2. IEDI: International Estimated Daily Intakes
3. STMR: Supervised Trials Median Residue
4. ADI: Acceptable Daily Intake
5. LOQ: Limit of quantification
6. The IEDI of other pesticides are available in Pesticide residues in food – 2005, *http://www.fao.org*

Source: Pesticide residues in food – 2005, http://www.fao.org/ag/AGP/AGPP/Pesticid/JMPR/DOWNLOAD/2005_rep/report2005jmpr.pdf

The above tables show that when people in different regions consume more of fruits and vegetables (shown by the Diet figures column) because of their inherited food culture or because of greater availability or because of greater awareness that fruits and vegetables are 'nutritious, hence good food', they are also taking more of commonly used pesticides. The Intake figures column shows a far higher intake of Acephate for Europeans than that for Africans or Far Easterners. Similarly, an European just by consuming more potatoes because of their inherited diet (who can stay away from a quick bite of tempting English fish 'n' chips) lands up taking in higher amounts of Chlorpropham, than that of Africans or Far Easterners.

Concluding Remarks

What is noteworthy that an awareness raising exercise is that apples, peaches, lettuce and many other 'good food' that we consume and give to our children have traces of many pesticides. What we have not been told very explicitly and widely by well meaning(?) NGOs or the Government in India is how much of these apples and peaches and lettuce etc are a pot-pourri of Cyhexatin, Indoxacarb and other residues of pesticides and agrochemicals. We understand that today's apples are not exactly what Adam & Eve had. What we don't understand is that instead of creating a great ruckus on pesticides in soft drinks in India and grab eye-ball attention by raving and ranting against MNCs in the electronic media, isn't it far better in the light of the FAO Report, 2005, to put down before the ordinary Indian readers credible recommendations made by Joint FAO/WHO Meeting on Pesticide Residues in fruits and vegetables and other staple food. Such recommendations are undertaken jointly by experts to evaluate possible hazards to humans arising from the occurrence of residues of pesticides. FAO and WHO put great effort into translating the outcome of risk assessments made by expert bodies into recommendations to experts in other areas and to enforcement bodies. We have the right to know more about, how safe are our basket of 'good foods'? As parents we should know whether giving large helpings of Indian apples or generous amount of rice-pudding accompanied by chunks of melting mangoes is really better for our children or is it safer to pop vitamin pills (at least once a while), washed down with red wine or honey or an occasional Coke & Pepsi. Surely good food has to be safe too. In the name of nutrition how much of pesticides are we being forced to guzzle down only because we don't have MRLs & ADIs for most pesticides used in our staple diet?

Endnotes

1 Pesticide residues in food – 2005. Report of the Joint Meeting of the FAO Panel of Experts on Pesticide Residues in Food and the Environment and the WHO Core Assessment Group on Pesticide Residues Geneva, Switzerland, 20-29 September 2005.

2 Pesticide Use: Now and then, *http://www.colorado.edu/MCEN/EnvTox/food.pdf*

3 Poison vs Nutrition, *http://www.cseindia.org/misc/cola-indepth/poison.pdf*

4 Food Safety Improvements Underway in China by Linda Calvin, Fred Gale, Dinghuan Hu, and Bryan Lohmar, Amber Waves, Nov 2006.

5 Indian NGO Finds Pesticides in Colas, *http://www.corpwatch.org/article.php?id=9668*

6 Ibid.

7 Ibid.

8 Pesticide Use: Now and then, *http://www.colorado.edu/MCEN/EnvTox/food.pdf*

9 Indian NGO Finds Pesticides in Colas, *http://www.corpwatch.org/article.php?id=9668*

10 Pesticide residues in food – 2005. Report of the Joint Meeting of the FAO Panel of Experts on Pesticide Residues in Food and the Environment and the WHO Core Assessment Group on Pesticide Residues Geneva, Switzerland, 20-29 September 2005.

11 Ibid.

12 Ibid.

Section III

Remedial Measures

7

Recommendations for Future Action*

D Atkinson, F Burnett, G N Foster, A Litterick, M Mullay and C A Watson

'Recommendations for future action', excerpted from the report 'The Minimisation Of Pesticide Residues in Food: A review of the published literature' is a review of publicly available information on the residues found in fresh foods currently being consumed in UK and of the scientific literature dealing with pesticide application. Data from a range of sources indicate that much of the food currently consumed in the UK contains residues of pesticides.

It is recommended that pesticide residues in foods might be reduced by obviating the need for pesticide applications through the adoption of ecological methods of agriculture or by the use of modern biotechnological means of crop enhancement so that chemical crop protection is no longer required. In the shorter term, progress can be made through changes in farming practice, the continuation of conventional plant breeding, through modification of current systems of crop protection with chemicals and through changes in storage practices. 'Recommendations for future action' gives a summary of potential actions to reduce selected residues in key crops.

* The article is based on the report "The Minimisation of Pesticide Residues in Food: A Review of the Published Literature".

1. General Conclusions

A review of publicly available information on the residues found in fresh foods currently being consumed in UK and of the scientific literature dealing with pesticide application and the fate of pesticides within the food chain has allowed a number of conclusions about how pesticide residues in foods might be reduced. Reducing residues in foods may be achieved by obviating the need for pesticide applications through the adoption of ecological methods of agriculture or by the use of modern biotechnological means of crop enhancement so that chemical crop protection is no longer required. Both of these are longer-term strategies, for reasons discussed in the body of the report. In the shorter term progress can be made through the continuation of conventional plant breeding, through modification of current systems of crop protection with chemicals and through changes in storage practices. The review of the scientific literature has lead to a series of conclusions on each of these issues which are presented at the end of each section. They are reproduced in abbreviated form at the end of this section (Table 1). A number of general conclusions about means of reducing pesticide residues in food can be drawn. These can be summarised as follows:

a) Data from a range of sources indicate that much of the food currently consumed in the UK contains residues of pesticides. Few of the current residues being detected by government, or those supermarkets that make their date publicly available, exceed MRLs. Despite the absence of a substantial proportion of samples exceeding the MRL application of the precautionary principal suggests the importance of using all reasonable means to minimise the quantity of pesticides found in the diet and the frequency with which individual foods are found to contain residues. While much may be achieved using current technologies and materials virtual elimination of residues would require a major rethink of current agricultural and horticultural systems which have been designed on the basis that crop protection would be delivered through chemicals.

b) Pesticide residues can however be reduced through changes in farming practice. Data from USA shows that residues in foods derived from organic systems are less than those found in ICM systems which, in turn, are less than those in food products from conventional systems where around 70% of samples contain some pesticide residue. The reduction in residues achieved in ICM systems, essentially a 50% reduction in the proportion of samples with residues, indicate the extent to which residues can be reduced in conventional agriculture by the application of currently available technologies. The presence of long-lived residues of older chemical in the soil may limit the ability of organic agriculture to deliver zero residue foods. Reducing pesticide residues in food is not, however, the only target for agriculture. Agriculture must also produce significant quantities of food and at a price the consumer will pay, and produce food of a quantity, including the absence of bruises and other external damage, which the consumer currently expects. Changes in farming practice to deliver lower residue food must be in balance with these other requirements.

c) Current developments in crop breeding should however both aid the productivity of organic systems and help reduce the need for pesticide use in conventional systems. The role of GM crops in reducing pesticides use in conventional systems remains equivocal. Its potential to reduce residues will depend upon whether acceptable weed control can be achieved in practice with broad-spectrum herbicides alone and whether resistance to pests, induced by GM technology, remain durable. GM technology appears to have the potential to enhance disease resistance in stored products and to modify physiology which would reduce the need for the categories of chemicals most often resulting in residues in stored products.

d) On the basis that residues are most commonly found in foods from conventional systems they have the greatest scope for reducing pesticides use and thus residues in food products. A deliberate planned strategy to reduce residues would be aided by information which allows the specifics of the use of individual pesticides to be linked to their occurrence in foods. Currently government surveys of pesticide use indicate best estimates of pesticide use on major crops. PRC data indicates the occurrence of some, but not all residues in major crops. Residue data indicates that some samples of crops contain residues while others do not. On the basis of current data it is not possible to explain why particular samples do or do not contain residues. If the delivery of fresh food products with low (no) residues were to become a design feature of agricultural systems then clear information chains which allowed use to be related to residues would be necessary. Currently UK data to parallel that from the USA presented by Baker *et al,* 2002, on the impact of major systems of production e.g. conventional, ICM, organic, is not publicly available.

e) Much of the fresh food consumed in the UK has been imported. PRC data suggests that currently imported food is slightly more likely to contain residues than that known to have been produced in the UK. Most of the conclusions we have drawn from the published literature on minimisation of residues are appropriate to both overseas producers and UK producers.

f) Reducing residues will not be achieved just by changes in pesticide use. The need for pesticides is a function of the distance foods are moved from production to consumption, the time which elapses between production and consumption and the desire of consumers to have available to them food products throughout the year. Changes in attitude to year long availability would aid a reduction in the need for pesticide use. Although information in the published literature on bruising and other change is relatively infrequent softening of consumer desires for blemish free produce would seem likely to enhance the scope for systems with reduced pesticide use.

g) Detailed conclusions from earlier sections are summarised in Table 1.

h) Reducing the total impact of residues could be aided by a series of specific actions related to those crops which are most often a source of significant residues. Potential actions are summarised in Table 2. Most of these actions could have benefit in the short term. Some of these are further developed in Section 2. Longer term needs are summarised in (i) below.

i) The options which seem likely to have the greatest impact on pesticide residues, based on our literature review, would seem to be:

   i. Maintaining breeding programmes, both conventional and GM, to increase and develop resistance to the main pests and diseases which are compatible with natural antagonists/symbionts.

   ii. To further develop understanding of the links between pesticide applications and the presence of residues in foods and to make available appropriate information.

iii. Development of organic agricultural and other ecological methods for those crops where they are most likely to be successful in producing substantial yields, quality and level of production.

iv. Transfer of methods of crop protection developed in organic farming to conventional and IPM systems.

v. Post harvest storage.

Table 1: A Summary of the Conclusions Drawn from the Literature Review

1. Pesticide Residues

(a) Losses of crop due to fungal diseases, pests and weeds remain significant considerations.

(b) Variation in the % of crop samples containing pesticide residues and in the number of residues found suggest scope to reduce residue levels and the % of samples with residues.

(c) Much current variation results from application practice.

(d) A high proportion of current residues result from post-harvest applications of growth regulators and fungicides.

(e) Fungicides were the most common pre-harvest materials found as residues.

(f) The availability of residue data for products produced by different agricultural systems is essential to assessing the impact of changes in agricultural husbandry on the generation of residues.

(g) Those involved in developing pesticide application strategies need to be more aware of the links between application practice and residues in foods.

(h) Data on pesticide residue levels for an increased number of commodities would be of value.

(i) The establishment of priorities must be informed by what is known of the residue levels through the food chain.

2. Organic Agriculture

(a) Pesticide residues derive from two principal sources:

(i) Soil residues of older chemicals.

(ii) Residues of other materials.

(b) As food from organic agriculture has reduced pesticide residues developing the scale of organic production would result in a reduction in the total amount of pesticide residues in foods.

(c) Increased investment in R & D should therefore increase yields and improve current crop protection practices.

(d) Currently there is a lack of robust scientific information comparing pesticide residue levels in organic and conventional foods.

Contd...

Contd...

(e) Most of the scientific reports on organic food quality have thus focused on differences between conventional and organic produce in terms of nutritional factors. Information on size, shape and appearance can generally only be inferred from percentage of outgrades of fruit and vegetables.

(f) Studies of pesticide residues in food could usefully include information on residues of potentially harmful botanical pesticides.

(g) There is a need to develop varieties suitable for organic growing conditions.

(h) Encouraging the growth of organic produce that is marketed fresh.

(i) Information on the yields of different crops and the ease of crop protection indicates that some crops could move to organic production more easily than others.

(j) The development of systems around chemical use has resulted in a severe narrowing of crop protection options. Conventional crop production could thus benefit from an increase in the range of options used. Drawing some areas of best practice from organic practice could aid in this process.

(k) There is a lack of published information on the effect of systems, which do not use pesticides on product quality in terms of shape, rise and appearance.

3. Biotechnological Agriculture

(a) Where GM has given crops resistance to pests, then the use of chemicals for pest control has generally fallen.

(b) GM crops have had little impact on fungicide use.

(c) Where GM crops have been developed for resistance to specific broad-spectrum herbicides then the use of that chemical has increased greatly.

(d) Where total herbicide use has increased this has usually given rise to residues in food which are potentially less toxic than are residues of other materials.

(e) If the aim of the GM modification is to reduce all residues in food then herbicide tolerant crops are unlikely to deliver this.

(f) The introduction of Ht varieties will significantly change systems of production.

4. Plant Breeding

(a) Breeding crops for resistance to pests and disease allows reductions in chemical use.

(b) Significant research activity on plant breeding will need to be maintained.

(c) Plant breeding for resistance to fungal diseases is likely to be an ongoing activity.

(d) Breeding for changes in growth which would obviate the need for sprout suppressants would be of significant value.

(e) Breeding gains can be additive.

(f) Breeding and associated approaches are unlikely wholly to eliminate the impact of pests or diseases.

Contd...

Contd...

5. Pesticide Application

(a) Past practice in the UK and elsewhere indicate that rates and amounts of pesticides applied to UK crops can be reduced.

(b) The UK climate makes reducing pesticide application more difficult than is the case for some other countries.

(c) ICM systems have been able to reduce both chemical applications and pesticide residues.

(d) Many of the newer chemicals e.g. strobilurin fungicides have been designed to be applied as prophylactic treatments. The philosophy of before the event protective sprays will limit the ability to make major reduction in residues of such materials.

(e) Weed control in the absence of chemicals has always been difficult. There is scope for disconnecting the link between application rates and residues.

(f) Reducing the use of chemicals in current systems will require a degree of redesign of systems.

6. Post-Harvest Pesticide Use

(a) Post harvest treatments are applied most directly to potential food products and close to the time of consumption. Minimising such applications will have a major impact on total residues.

(b) The contribution of post harvest materials can be reduced by changes in consumer habits.

(c) A reduction in post-harvest pesticide use without alternative action would result in increased insect and mite infestations.

(d) Reduced pesticide usage would be aided better training of store managers, more enforcement of basic management standards, more widespread installation of ventilation systems, better design of ventilation systems and higher specifications for grain stores.

(e) Post-harvest residues make up a significant proportion of the sum of all residues found and so action through improved store design and management and would be likely to have a major impact on residues found.

Table 2: A Summary of Potential Actions to Reduce Selected Residues in Key Crops

Action/Potential Action	Comment	Reference or Potential Contact
TOP FRUIT – Fungicides		
Incorporation of resistance to powdery mildew in many cvs	Acceptable season of harvest, storage period, shelf life and flavour usually take priority in breeding programmes	Alston *et al*, 1988

Contd...

Contd...		
Choice of treatment of apple scab	The weighting of the criteria for comparison, cost, residue potential, compatibility with biological control, affects the choice	Kovach *et al*, 1992
Half of detected pesticide residues result from post-harvest treatment outwith the UK.	Post harvest materials are applied close to the point of consumption and are less exposed to rain, thus generating most residues.	Storage Sector
Increasing usage of strobilurins	May reduce inputs of pesticide-generating fungicides	Clough and Godfrey, 1998
SOFT FRUIT – Fungicides		
Fungicide usage high for grey mould	What are viable alternatives? Cultural control measures increasing air flow around plants can be beneficial. Biological control measures have also been demonstrated to have efficacy.	Larkin *et al*, 1998
Post-harvest treatments appear to be increasing	The increasing role of strobilurins. Grey mould has a history of developing resistance fast and strobilurins are not likely to be effective for very long in this scenario.	Syngenta, BASF
TOMATOES – Fungicides		
Mixed cultivar cropping	Practised for top fruit and cereals so potential for greenhouse crops? Main concern is grey mould?	Horticultural Industry, SAC
Choice of fungicide can affect success of biological control	Integrated advice on choice of all management treatments crucial for pest management	Biocontrol agent suppliers
Soil-less systems have massively reduced soil-borne pathogens		Van Heyningen Brothers
CELERY and LETTUCE – Pesticide Input		
Use of propyzamide as pre-emergence herbicide	Possible substitution of mulches	Davies, SAC
		Contd...

Contd...		
Aphicides for root aphids	Some resistant cultivars available, also irrigation might help	Chemical Industry
Foliar pests and diseases	Successional plantings cause problems to build up	Chemical Industry
POTATOES – Storage		
Reduce CIPC usage by a low-dose programme	*Luxan Gro-Stop HN* and *Gro-Stop Fog* are applied at 18 g a.i. per tonne per season. This should reduce CIPC residues. Low-dose fogging programme is at lower temperature and over short period than high dose, minimising disturbance to store temperature	Malcolm Nursey, Luxan
Improve uniformity of CIPC application by better understanding of physics of deposition	Further research required to understand the physics of deposition of CIPC and to evaluate the effect of air distribution, condensation, and benefit from positive ventilation through the potatoes	Miller *et al* (2003).
Keep potatoes dry and free from condensation, and at a low temperature	Requires investment in new and up-graded storage facilities and adoption of new environmental control algorithms	Pringle *et al* (2001).
GRAIN – Storage Pesticides		
Reduce pesticide input by installing better environmental equipment and controls	Capital cost perceived to be prohibitive but cost effective in the long term	McGovern, SAC
GRAIN – Effects of Prophylactics		
Decision support to minimise strobilurin input, thus reducing need for glyphosate pre-harvest	Benefits of strobilurin treatment reflected in yield. Fungicide resistance in key cereal pathogens may limit the usage of this group, especially late in the season	Burnett, SAC
Seek alternative desiccants to glyphosate		Chemical Industry

2. Recommendations for Future Action

A review of the published scientific literature indicates that pesticide residues are found at some level in a significant proportion of the food production consumed in the UK. It shows that the use of current technologies, appropriately employed, have scope to lead to a reduction in residues. Through the medium of the review of published literature we have indicated a number of approaches which might be followed. For approaches to be followed they would need to be practical and economically viable. Establishing which of the options we identify from our review of the literature might lead to a practical impact upon pesticide residues would benefit from a series of conversation with key stakeholders. Stakeholders conversations within a sector would lead to a report which identified the initial response of stakeholders within that sector and identified the series of next steps.

2.1 The principle options suggested by published literature as means of reducing pesticides are as shown in Figure 1.

Figure 1: Options for Reducing Pesticide Residues in Foods and their Links to Stakeholder Communities

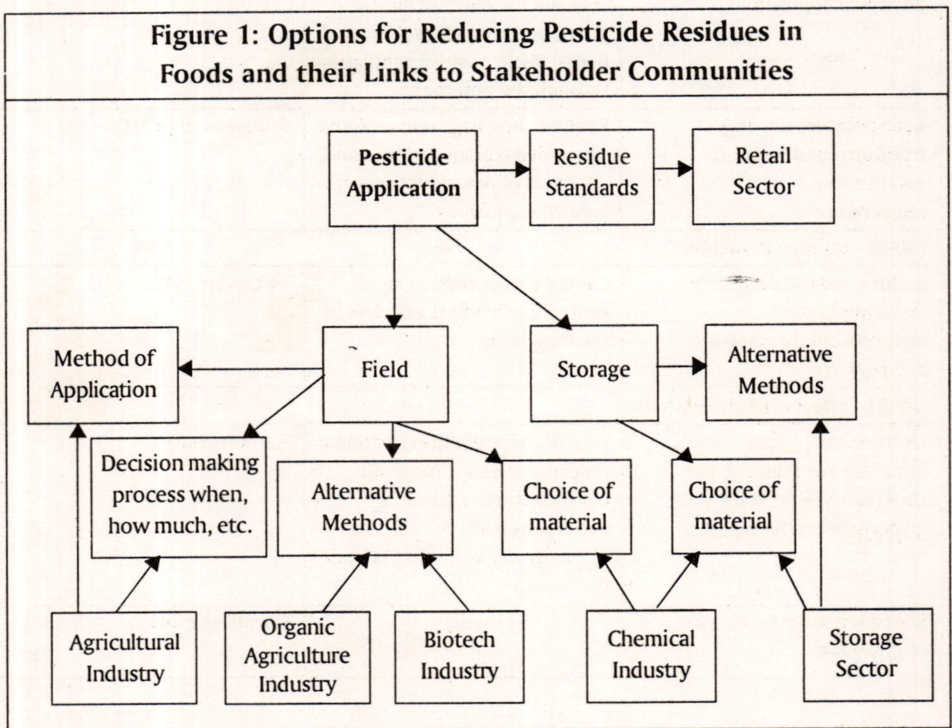

This suggests the following stakeholder conversation as important:

2.1.1 The Agricultural Industry

The farming industry selects the pesticides, which are to be applied, decides whether, when and how often they should be applied and makes the application. The review indicates that residue levels are influenced by all of these factors. The review also indicates that similar crop protection solutions can lead to different pesticide residue outcomes; even though the impact of different elements of the decision chain is unclear.

The review suggests that because current usage of pesticides is a consequence of the design of current agricultural and horticultural production systems that major changes may require a rethink of some of the more basic aspects of system. It would need industry action and acceptance for any wide scale changes to current systems.

A major stakeholder group would therefore be the agricultural community. Initial conversations would need to include leading agricultural and horticultural producers and representatives of organisations such as LEAF.

2.1.2 Chemical Industry

The materials available for use by farmers and growers are a function of those produced by chemical industry. Published literature suggests that while chemical industry has long been aware of the need for pesticide to have low human toxicity and more recently minimal impact on the environment and on non-target organisms chemicals do not seem to have been explicitly selected for their ability not to result in residues. This may be a consequence of the primary aim of producing chemicals with low human toxicity. The presence of residues in foods is a result of the structure of the chemical which will influence its degradation within the crop, its movement to and degradation within the food product, its application characteristics, i.e. for fungicides whether it is applied prophylactically or following infection, and the rate and frequency at which it needs to be applied. Were systems to be designed to deliver minimised levels of pesticides then design of chemical which produced low residues would be important. Although most chemical manufacturers are based in Europe or USA most have significant UK offices with which a conversation of this type could be carried out. Such discussions with chemical industry would include chemicals applied both pre- and post-harvest.

2.1.3 Biotechnology Industry

The study identified that developments in crop genetics had the potential to both reduce and increase chemical use and the use of particular chemicals. Reductions in chemical use would be a function of genes for resistance to pests and diseases being incorporated into the plant genome. Other changes i.e. genes for herbicides resistance, might increase herbicide use in total and the use of some specific herbicides. Similarly incorporation of genes allowing crops to be grown on poorer classes of land, which currently tends to be grassland, could lead to increased chemical use as a consequence of the changed land use. The biotechnology industry has recently established an umbrella body, the Agricultural Biotechnology Council that includes participation from BASF, Bayer, Dow, DuPont, Monsanto and Syngenta. A discussion with this group and with others involved in biotechnology, including academic organisations, would be helpful. Discussions might usefully include options for reducing post harvest growth regulators by crop engineering.

2.1.4 The Alternative Agriculture Sector

The review of the currently available literature identified that levels of currently used pesticides were low in organic produce. A move to organic production would therefore reduce pesticides in foods. This change would also however have significant impact on the volume of crop production, the quality and appearance of food produced and would have major impact on the structure of the current farming industry; most current organic farms are small and moving to a small farm model at a time when the average size of production units in most major producer countries is changing in the reverse direction and increasing in size would therefore involve major structural changes to the agricultural industry.

In the absence of major changes in farm structure the literature suggests that some current practices used on organic farms could be used both in ICM systems and within the conventional farm sector. In addition there may be scope for some horticultural commodities which current studies show frequently contain pesticide residues to be produced using organic methods as part of a total plan to reduce the overall quality of residues in the diet. Discussions with stakeholders in the organic and relevant sectors would help in developing these options and allow exploration of the scope for reducing imperfections in fresh products, which are sometimes found in such systems.

2.1.5 The Storage Sector

The literature review identified the post-harvest (storage) sector as the largest single generator of pesticide residues. The use of pesticides includes both growth regulators to arrest physiological development in stored products and the use of fungicides and insecticides to reduce spoilage. Changes to the storage environment and changes in the length of time that produce is stored could influence the need to use pesticides. Increased training for those involved in the operation of stores, may also be an area for action. A discussion with major store operators, especially those currently storing potatoes and top fruit (apples, citrus) should help scope what might be achieved through the application of currently available technology.

2.1.6 The Supermarket Sector

The retail sectors perception of pesticide use and pesticide residues and the information which they transmit to customers is a major element in the wider pesticide debate. Major supermarkets currently have good databases of information on residues, although only Coop and M&S make it publicly available. Being able to link pesticide use on the farm (or post harvest) and the occurrence of residues is a key element in devising strategies which would either reduce the scope of the pesticide use—pesticide residue link or decouple the use—residue linkage so allowing use without the production of residues. Conversations with those retail chains who currently make information available and those who don't would help develop the wider picture.

2.2 Outcome

The aim of the stakeholder for these listed above would be to produce a report detailing the issues covered in discussions, and especially the industry sectors reaction to the options identified by the literature review, and on the basis of these discussions to suggest both how a wider of series stakeholder discussions and other actions might, most helpfully, be taken ahead.

(Atkinson D, SAC – Edinburgh, The Scottish Agricultural College (SAC), United Kingdom,

F Burnett, Researcher (Plant Pathology), Crop & Soil Systems Research Group King's Buildings, Edinburgh,

G N Foster, can be reached at g.foster@au.sac.ac.uk,

A Litterick, Environmental Consultant, Ferguson Building, Craibstone Estate, Aberdeen,

M Mullay, Web Editor, SAC, West Mains Road, Edinburgh,

C A Watson, SAC – Aberdeen, The Scottish Agricultural College (SAC), United Kingdom.)

APPENDIX

Pesticide Use

Table A: Pesticide Usage on Pre-Harvest Cereals in Britain

(tonnes active ingredient)

	Insecticides	Molluscicides	Fungicides	Growth Regulators	Seed Treatments	Herbicides	Total
1988	292	138	4,458	2,034	168	10,193	17,115
1994	406	179	2,733	2,476	404	5,308	11,104
1996	176	97	3,145	2,774	284	6,547	12,740
1998	223	95	2,630	3,001	265	6,871	12,729

Table B: Pesticide Usage on Pre-Harvest Potatoes in Britain

(tonnes active ingredient)

	Insecticides Nematicides	Molluscicides	Fungicides	Growth Regulators	Seed Treatments	Herbicides	Desiccants	Total*
1988	122	11	1,000	3	58	259	5,418	1,453
1994	115	13	1,158	72	46	266	12,806	1,670
1996	454	18	1,135	37	48	313	13,113	2,005
1998	151	26	1,339	31	42	264	12,727	1,853

* excluding desiccants, dominated by sulphuric acid.

Table C: Pesticide Usage on British Tomato, Cucumber and Pepper Crops

(kg)

	Area (ha)	Area with b.c. (ha)	Insecticides#	Mollus -cicides	Fungicides~	Herbicides	Soil Sterilants*	Total
1991	749	16,660	2,228	9	17,763	24	47,565	20,283
1995	581	13,859	1,268	14	17,600	15	32,813	18,912
1999	572	9,816	941	2	6,038	2	14,868	6,983

excludes registered biological control products, which are included under areas treated with other biological control agents

~ includes sulphur, increasing as preferred usage in presence of biological control agents

* estimates of soil sterilants variable, declining through switch to soil-less cultivation

Table D: Pesticide Usage on British Tomato Crops

(kg)

	Area (ha)	Area with b.c.(ha)	Insecticides#	Mollus -cicides	Fungicides~	Herbicides	Soil Sterilants*	Total
1991	444	11,731	811	5	13,444	14	44,590	14,274
1995	347	10,350	863	8	8,807	15	30,489	9,693
1999	295	6,159	579	<1	4,560	2	12,063	5,141

\# excludes registered biological control products, which are included under areas treated with other biological control agents

~ 6% (1991), 22% (1995), 32% (1995) by weight sulphur used for powdery mildew control in conjunction with biological control of pests

* estimates of soil sterilants variable, largest values due to detection of use of disinfectants.

Table E: Pesticide Usage on British Cucumber Crops

(kg)

	Area (ha)	Area with b.c.(ha)	Insecticides#	Molluscicides	Fungicides	Herbicides	Soil Sterilants*	Total
1991	239	4,104	1,332	–	4,208	1	984	5,541
1995	181	2,653	289	1	8,655	–	2,324	8,945
1999	244	3,524	347	2	1,478	–	2,805	1,827

\# excludes registered biological control products, which are included under areas treated with other biological control agents

* estimates of soil sterilants variable, largest values due to detection of use of disinfectants.

Table F: Pesticide Usage on British Pepper Crops

(kg)

	Area (ha)	Area with b.c.(ha)	Insecticides#	Molluscicides	Fungicides	Herbicides	Soil Sterilants	Total*
1991	66	825	85	4	111	9	1,991	468
1995	53	856	116	5	138	–	–	274
1999	33	133	15	–	–	–	–	15

\# excludes registered biological control products, which are included under areas treated with other biological control agents

* estimates of soil sterilants variable, any trend being the result of move to soil-less cultivation.

Table G: Pesticide Usage on British Lettuce Crops

(kg)

	Area (ha)	Insecticides	Molluscicides	Fungicides	Herbicides	Soil Sterilants	Total*
1991	1,042	473	64	11,685	360	58,055	12,582
1995	700	312	67	17,763	187	14,007	18,329
1999	481	326	13	7,983	120	25,152	8,442

* estimates of soil sterilants variable, 1% of area but 75% of weight in 1999.

Table H: Pesticide Usage on British Celery Crops

(kg)

	Area (ha)	Insecticides	Molluscicides	Fungicides	Herbicides	Soil Sterilants	Total*
1991	101	52	24	491	60	2,997	627
1995	103	46	12	254	71	–	383.
1999	50	31	5	282	18	1,975	336

* excludes unreliable estimates of soil sterilants, 1% of area but 83% of weight in 1999.

Table I: Pesticide Tonnages in Key Crops in Scotland

	Cereals			Potatoes				Ware potatoes in store# GR/F	
	F	I/M	H/GR	F	I/M	H/GR	SA	Merchant	Farm
2000 *f*	303	19	601	208	16	38	5,552		
1998*	508	19	649	185	17	42	6,122	1.9	3.4
1996**	483	9	635	156	12	46	6,038	0.5	5.8
1994**	336	14	567	157	12	52	5,356	0.2	4.8
1992**	515	11	668	144	14	3,639			

F = fungicides; GR = growth regulators; H = herbicides; I = insecticides; M = molluscicides; SA = sulphuric acid

* Snowden & Thomas, 1999; ** Snowden & McCreath, 1997; # Thomas & Snowden 2001; *f* Kerr & Snowden 2001.

Table J: Pesticide Tonnage's in Key Crops in Scotland

	Soft Fruit			Vegetables		
	F	I/M	H	F	I/M	H
2001*f*	13.8	0.9	3.8			
1999#				12.1	8.5	25.3
1998*	20.8	5.2	8.1			
1995#				17.5	14.1	42.8
1994*	19.1	5.5	6.1			
1991#				19.2	13.2	33.1
1990*	18.9	5.3	5.8			
1989##				11.9	11.3	32.8
1986**	14.3	4.7	10.3			

* McCreath 2000

** McCreath & Snowden 1994.

\# Snowden & Thomas, 2000

\## McCreath 1996.

f Snowden 2002.

Glossary of Terms Used

a.i.	Active Ingredient
ACP	Advisory Committee on Pesticides
ADI	Accepted Daily Intake
AMPA	Amino Methylinphosphonic Acid
ARfD	Acute Reference Dose
BCPC	British Crop Protection Council (an umbrella body of public and private organisations concerned with crop protection)
Bt	*Bacillus thuringiensis* (the source of genes for insect resistance introduced into crops)
CIPC	Chorpropham
DDT	Dichloro-diphenyl trichloro-ethane
DSS	Decision Support System
EIQ	Environmental Impact Quotient
EPA	Environmental Protection Agency USA
FQPA	Food Quality Protection Act USA

Contd...

Contd...	
GAP	Good Agricultural Practice
GC	Gas Chromatography
GM	Genetically Modified
GMO	Genetically Modified Organism
HPLC	Higher Pressure Liquid Chromatography
HR	Hypersensitive Reaction
Ht	Herbicide Tolerant (Genetically Modified Crops)
ICM	Integrated Crop Management (a system integrating a range of methods used for both crop production and protection)
IFOAM	International Federation of Organic Agricultural Movements
IPM	Integrated Pest Management (a system involving a series of pest management methods)
LOD	Limit of Detection
MRL	Maximum Residue Level
NIAB	National Institute of Agricultural Botany
NOAEL	No Observable Adverse Effect Level
OP	Organo Phosphorus Insecticides
OSR	Oil Seed Rape
PCB	Poly Chloro biphenol
PRC	Pesticides Residues Committee
PRI	Potential for Residues Index
PSD	Pesticide Registration Directorate
PSPS	Pesticide Safety Precaution Scheme
RR	Round-up Ready (crops resistant to the herbicide glyphosate)
SA	Soil Association
SAC	Scottish Agricultural College
SAR	Systemic Acquired Resistance
TBZ	Thiabendazole
UKROFS	United Kingdom Register of Organic Food Standards
USDA	United States Department of Agriculture
WiGRAMP	Working Group for the Risk Assessment of Mixtures of Pesticides
WPPR	Working Party on Pesticide Residues

8

Organic Farming: Its Relevance to the Indian Context

P Ramesh, Mohan Singh and A Subba Rao

Increasing consciousness about conservation of environment as well as health hazards associated with agrochemicals and consumers' preference to safe and hazard-free food are the major factors that lead to the growing interest in alternate forms of agriculture in the world. Organic agriculture is one among the broad spectrum of production methods that are supportive of the environment. The demand for organic food is steadily increasing both in the developed and developing countries with an annual average growth rate of 20–25%. Organic agriculture, without doubt, is one of the fastest-growing sectors of agricultural production. However, there are certain issues that should be clarified before we go for a large-scale conversion to organic agriculture. The most important issues are—Can organic farming produce enough food for everybody? Is it possible to meet the nutrient requirements of crops entirely from organic sources? Are there any significant environmental benefits that accrue from organic farming? Is the food produced by organic farming superior in quality? Is it economically feasible? In this

Source: www.ias.ac.in, Indian Academy of Sciences, Current Science, Vol. 88, No. 4, 25 February 2005.

article, we review these aspects of organic farming. In India, vast stretches of arable land, which are mainly rain-fed and found in the Northeastern region where negligible amount of fertilizers and pesticides are being used and have low productivity, could be exploited as potential areas for organic agriculture. Considering the potential environmental benefits of organic production and its compatibility with integrated agricultural approaches to rural development, organic agriculture may be considered as a development vehicle for developing countries like India, in particular.

Green revolution technologies involving greater use of synthetic agrochemicals such as fertilizers and pesticides with adoption of nutrient-responsive, high-yielding varieties of crops have boosted the production output per hectare in most cases. However, this increase in production has slowed down and in some cases there are indications of decline in productivity and production. Moreover, the success of industrial agriculture and the green revolution in recent decades has often masked significant externalities, affecting natural resources and human health as well as agriculture itself[1]. Environmental and health problems associated with agriculture have been increasingly well documented, but it is only recently that the scale of the costs has attracted the attention of planners and scientists. The external costs of agriculture in the UK have been estimated as 1.1–3.9 billion pounds per annum[2]. As the external costs of farming are not internalized in the price of food, tax payers (or more likely the future generations) will have to pay the bill that is getting bigger every day.

Increasing consciousness about conservation of environment as well as of health hazards caused by agrochemicals has brought a major shift in consumer preference towards food quality, particularly in the developed countries. Global consumers are increasingly looking forward to organic food that is considered safe and hazard-free. The global market for organic food is expected to touch US$29 to 31 billion by 2005. The demand for organic food is steadily increasing both in developed and developing countries, with annual average growth rate of 20-25%. Worldwide, over 130 countries produce certified organic products in commercial quantities[3].

The Concept of Organic Agriculture

Organic agriculture is one among the broad spectrum of production methods that are supportive of the environment. Organic production systems are based on specific standards precisely formulated for food production and aim at achieving agro ecosystems, which are socially and ecologically sustainable. It is based on minimizing the use of external inputs through use of on-farm resources efficiently compared to industrial agriculture. Thus the use of synthetic fertilizers and pesticides is avoided.

Organic Farm at IISS, Bhopal

Leaf Litter in Pigeonpea Crop – A Source of Organic Manure

'Organic' in organic agriculture is a labelling term that denotes products that have been produced in accordance with certain standards during food production, handling, processing and marketing stages, and certified by a duly constituted certification body or authority. The organic label is therefore a process claim rather than a product claim. It should not necessarily be interpreted to mean that the foods produced are healthier, safer or all natural. It simply means that the products follow the defined standard of production and handling, although surveys indicate that consumers consider the organic label as an indication of purity and careful handling. Organic standard will not exempt producers and processors from compliance with general regularity requirements such as food safety regulations, pesticide registrations, general food and nutrition labelling rules, etc.[4]

Many definitions have been proposed for organic agriculture. Ethical issues such as fair labour practices and animal ethics have also been included in organic agriculture definitions[5]. All these however primarily focus on ecological principles as the basis for crop production and animal husbandry. To promote organic agriculture and to ensure fair practices in international trade of organic food, the Codex Alimentarius Commission, a joint body of FAO/WHO framed certain guidelines for the production, processing, labelling and marketing of organically produced foods, with a view to facilitate trade and prevent misleading claims[6].

The Codex Alimentarius Commission defines organic agriculture as a holistic food production management system, which promotes and enhances agro ecosystem health, including biodiversity, biological cycles and soil biological activity. It emphasizes the use of management practices in preference to the use of off-farm inputs, taking into account that regional conditions require locally adapted systems. This is accomplished by using, where possible, agronomic, biological and mechanical methods, as opposed to using synthetic materials, to fulfil any specific function within the system.

Organic Versus Conventional Agriculture

In recent years, there is a lot of debate between the proponents of organic farming and a section of the community who questioned the scientific validity and feasibility of organic farming[7]. The most often debated issues on organic agriculture fall under the following six categories:

i) Can organic farming produce enough food for everybody?

ii) Is it possible to meet the nutrient requirements of crops entirely from organic sources?

iii) Are there any significant environmental benefits of organic farming?

iv) Is the food produced by organic farming superior in quality?

v) Is organic agriculture economically feasible?

vi) Is it possible to manage pests and diseases in organic farming?

In this article, a brief review on the above aspects of organic farming is presented. The review is made mostly from research work conducted elsewhere, since most of the published data on organic farming comes from developed nations; but wherever possible, Indian literature is taken into account. This review is not exhaustive, but indicative on certain pertinent aspects of organic farming, which are often talked about and debated, particularly with reference to its adoption among Indian farming communities, especially under rain-fed zones, tribal areas and northeast mountain areas, where use of agrochemical is low or negligible.

Can Organic Farming Produce Enough Food for Everybody?

Organic Agriculture and Yield

Yields relative to comparable conventional systems are directly related to the intensity of farming of the prevailing conventional systems. This is not only the case for comparison between regions, but also between crops within a region, and for individual crops over time[8]. An over-simplification of the impact of conversion to organic agriculture on yield indicates that:

i) In intensive farming systems, organic agriculture decreases yield; the range depends on the intensity of external input use before conversion[9–12].

ii) In the so-called green revolution areas (irrigated lands), conversion to organic agriculture usually leads to almost identical yields[13,14].

iii) In traditional rain-fed agriculture (with low external inputs), organic agriculture has shown the potential to increase yields[15,16].

A number of studies have shown that under drought conditions, crops in organic agriculture systems produce significantly higher yields than comparable conventional agricultural crops[9,17], often out-yielding conventional crops[10,18,19] by 7-90%. Others have shown that organic systems have less long-term yield variability[20,21]. A survey of 208 projects in developing tropical countries in which contemporary organic practices were introduced, showed average yield increases of 5-10% in irrigated crops and 50-100% in rainfed crops[22]. The so-called organic transition effect, in which a yield decline in the first 1-4 years of transition to organic agriculture, followed by a yield increase when soils have developed adequate biological activity[20,23,24], has not been borne out in some reviews of yield comparison studies[9,25]. Trials conducted on organic cotton at Nagpur indicated that after the third year, the organic plot, which did not receive fertilizers and insecticides, produced as much cotton as that cultivated with them[13]. Similarly, studies conducted in Punjab clearly indicated that organic farming gave higher or equal yields of different cropping systems compared to chemical farming after an initial period of three years[14].

Organic Agriculture and Food Security

The common claim that large-scale conversion to organic agriculture would result in drastic reduction in world food supplies or large increases in conversion of undisturbed lands to agriculture[26,27], has not been borne out in modelling studies. Conversion studies showed that domestic food consumption would not suffer, exports would vary depending on crop, but the structure of farming would definitely change with more diversification of agriculture[28]. Widespread conversion to organic agriculture would result in crop yield increase over the current averages as a result of increased investment in research and extension[29]. In northern Europe domestic food demand could be met with organic methods, but food would be more expensive[30]. Similar conclusions were also reached in studies in the US[31]. A German study concluded that if the per capita dietary calories from meat were reduced to 21% from the current 39%, all German crop lands could be converted to organic agriculture without an increase in imports or expansion of agricultural land. According to the report such a conversion would be possible by 2017, as the survey indicated that 40% of German youth have planned to maintain low or no-meat diets[32]. The threshold for this conversion may come sooner, as abandonment of meat-consumption is accelerating in Europe as a result of BSE[33]. Several studies have proved the benefits of vegetarian diet over non-vegetarian in terms of energy consumed for food production as well as other nutritional advantages. Organic farmers grow a variety of crops and maintain livestock in order to optimize use of nutrients and the space between species. This ensures economic advantages through low crop production or yield failure due to biotic and abiotic factors in all of these simultaneously. This can have an important impact on local food security and resilience. In rain-fed systems, organic agriculture has demonstrated to out-perform conventional agricultural systems under environmental stress conditions[9,10,20]. Under the right circumstances, the market returns from organic agriculture can potentially contribute to local food security by increasing family income. At the global level, especially in developing countries with high population pressure, and with the present state of knowledge and technology, organic farmers cannot produce enough food for everybody.

Nutrient Management in Organic Farming

Organic farming is often understood as a form of agriculture with use of only organic inputs for the supply of nutrients and management of pests and diseases. In fact, it is a specialized form of diversified agriculture, wherein problems of farming are managed using local resources alone. The term organic does not explicitly mean the type of inputs used; rather it refers to the concept of farm as an organism. Often, organic agriculture has been criticized on the grounds that with organic inputs alone, farm productivity and profitability might not be improved because the availability of organic sources is highly restricted[7]. True, organic resources availability is limited; but under conditions of soil constraints and climate beggaries, organic inputs use has proved more profitable compared to agrochemicals[15].

Organic farming systems rely on the management of soil organic matter to enhance the chemical, biological and physical properties of the soil. One of the basic principles of soil fertility management in organic systems is that plant nutrition depends on 'biologically-derived nutrients' instead of using readily soluble forms of nutrients; less available forms of nutrients such as those in bulky organic materials are used. This requires release of nutrients to the plant via the activity of soil microbes and soil animals. Improved soil biological activity is also known to play a key role in suppressing weeds, pests and diseases[34].

Animal dung, crop residues, green manure, biofertilizers and bio-solids from agro-industries and food processing wastes are some of the potential sources of nutrients of organic farming. While animal dung has competitive uses as fuel, it is extensively used in the form of farmyard manure. Development of several compost production technologies like vermicomposting, phosphocomposting, N-enriched phos-phocomposting, etc. improves the quality of composts through enrichment with nutrient-bearing minerals and other additives. These manures have the capacity to fulfil nutrient demand of crops adequately and promote the activity of beneficial macro- and micro-flora in the soil[35].

There are several doubts in the minds of not only farmers, but also scientists about whether it is possible to supply the minimum required nutrients to crops through organic sources alone. Even if it is possible, how are we going to mobilize the organic matter? At this juncture, it is neither advisable nor feasible to recommend the switchover from fertilizer use to organic manure under all agro-ecosystems. Presently, only 30% of our total cultivable areas has irrigation facilities where agrochemicals use is higher compared to rain-fed zones. It is here that ingenuity and efforts are required to increase crop productivity and farm production despite recurrence of environmental constraints of drought and water scarcity.

The basic requirement in organic farming is to increase input use efficiency at each step of the farm operations. This is achieved partly through reducing losses and adoption of new technologies for enrichment of nutrient content in manure[36]. Technologies to enrich the nutrient supply potential from manure, including farmyard manure three to four times are being widely used in organic farms. According to a conservative estimate, around 600 to 700 million tonnes (mt) of agricultural waste is available in the country every year, but most of it is not used properly. We must convert our filth into wealth by mobilizing all the biomass in the rural and urban areas into bioenergy to supply required nutrients to our starved soil and fuel to farmers[37]. India produces about 1800 mt of animal dung per annum. Even if two-thirds of the dung is used for biogas generation, it is expected to yield biogas not less than 120 mm^3 per day. In addition, the manure produced would be about 440mt per year[38], which is equivalent to 2.90mt N, 2.75mt P_2O_5 and 1.89mt K_2O.

Organic farms and food production systems are quite distinct from conventional farms in terms of nutrient management strategies. Organic systems adopt management options with the primary aim to develop whole farms, like a living organism with balanced growth, in both crops and livestock holding. Thus nutrient cycle is closed as far as possible. Only nutrients in the form of food are exported out of the farm. Crop residues burning is prohibited; so also the unscientific storage of animal wastes and its application in the fields. It is, therefore, considered more environment friendly and sustainable than the conventional system. Farm conversion from high-input, chemical-based system to organic system is designed after undertaking a constraint analysis for the farm with the primary aim to take advantage of local conditions and their interactions with farm activities, climate, soil and environment, so as to achieve (as far as possible) closed nutrient cycles with less dependence on off-farm inputs. This implies that the only nutrients leaving the farm unit are those for human consumption.

Crop rotations and varieties are selected to suit local conditions having the potential to sufficiently balance the nitrogen demand of crops. Requirements for phosphorus, sulphur and micronutrients are met with local, preferably renewable resources. Organic agriculture is, therefore, often termed as knowledge-based rather than input-based agriculture. Furthermore, organic farms aim to optimize the crop productivity under a given set of farm conditions. This is in contrast to concept of yield maximization through the intensive use of agrochemicals, irrigation water and other off-farm inputs. There are ample evidences to show that agrochemical-based, high-input agriculture is not sustainable for long periods due to gradual decline in factor productivity, with adverse impact on soil health and quality[1,8].

Environmental Benefits of Organic Agriculture

The impact of organic agriculture on natural resources favours interactions within the agro-ecosystem that are vital for both agricultural production and nature conservation. Ecological services derived include soil forming and conditioning, soil stabilization, waste recycling, carbon sequestration, nutrient cycling, predation, pollination and habitats[34].

The environmental costs of conventional agriculture are substantial, and the evidence for significant environmental amelioration via conversion to organic agriculture is overwhelming[39,40]. A review of over 300 published reports[41] showed that out of 18 environmental impact indicators (floral diversity, faunal diversity, habitat diversity, landscape, soil organic matter, soil biological activity, soil structure, soil erosion, nitrate leaching, pesticide residues, CO_2, N_2O, CH_4, NH_3, nutrient use, water use and energy use), organic farming systems performed significantly better in 12 and performed worse in none. There are also high pre-consumer human health costs to conventional agriculture, particularly in the use of pesticides[42]. It is estimated that 25 million agricultural workers in developing countries are poisoned each year by pesticides[43].

Safety and Quality of Organically Produced Food

There is a growing demand for organic foods driven primarily by the consumer's perceptions of the quality and safety of these foods and to the positive environmental impact of organic agriculture practices. The 'organic' label is not a health claim, it is a process claim. It has been demonstrated that organically produced foods have lower levels of pesticides and veterinary drug residues and in many cases lower nitrate contents. No clear trends have, however, been established in terms of organoleptic quality differences between organically and conventionally grown foods.

There have been many claims that eating organic foods increases exposure to microbiological contaminants[44]. But studies investigating these claims have no evidence to support them[45–48]. Organic foods must meet the same quality and safety standards applied to conventional foods. These include the CODEX General Principles of Food Hygiene and Food Safety Programmes based on the Hazard Analysis and Critical Control Point[6]. Analysis of pesticide residues in produce in the US and Europe has shown organic products have significantly lower pesticide residues than conventional products[4,49,50]. Nitrates are significant contaminants of foods, generally associated with intensive use of nitrogen fertilizers. Studies that compared nitrate contents of organic and conventional products found significantly higher nitrates in conventional products[4,49,51].

There are also claims that food produced by organic methods tastes better and contains a better balance of vitamins and minerals than conventionally grown food. However, there is no clear scientific evidence, with some studies showing increase in vitamin C, minerals and proteins[52], more sweeter and less tart apples[53] and others not[49]. A crude analysis of the literature, however, favours organic products in this area[54]. A tasting panel convened by the Consumer Association in the United Kingdom did not consistently favour the taste of organic fruits and vegetables[55]. Quality after storage has been reported to be better in organic products relative to conventional products after comparative tests[53,56]. Reviews of organic vs conventional product sensory analysis studies have reported results that do not clearly substantiate claims of superior organic product tastiness[49]. It is a known fact that the quality of crops is controlled by a complex interaction of factors, including soil type and the ratio of minerals in added compost, manure and fertilizer. So it is difficult to separate the influence of the environment and farming system[57]. There is scope to generate information on the quality of produce generated on organic farms in future studies.

Economics of Organic Farming

The replacement of external inputs by farm-derived resources normally leads to a reduction in variable input costs under organic management. Expenditure on fertilizers and sprays is substantially lower than in conventional systems in almost all the cases[12,30]. In a few cases, higher input costs due to the purchase of compost and other organic manures have been reported[58]. Studies have shown that the common organic agricultural combination of lower input costs and favourable price premiums can offset reduced yields and make organic farms equally and often more profitable than conventional farms[19,53,59]. Studies that did not include organic price premiums have given mixed results on profitability[60]. Studies from Europe and Canada show labour costs in organic agriculture average 40-50% higher where the wage rates are generally higher[30]. Gross margins, the difference between farm output and variable costs are generally similar or, where there are favourable price premiums, higher in organic agriculture. The economics of organic cotton cultivation over a period of six years indicated that there is a reduction in cost of cultivation and increased gross and net returns compared to conventional cotton cultivation in India[13].

Pest and Disease Management in Organic Farming

Pest control in organic farming begins by making sensible choices, such as growing crops that are naturally resistant to diseases and pests, or choosing sowing times that prevent pest and disease outbreaks. Careful management in both time and space of planting not only prevents pests, but also increases population of natural predators that can contribute to the control of insects, diseases and weeds[61]. Other methods generally employed for the management of pests and diseases are: improving soil health to resist soil pathogens and promote plant growth; rotating crops; encouraging natural biological agents for control of diseases, insects and weeds; using physical barriers for protection from insects, birds and animals; modifying habitat to encourage pollinators and natural enemies of pests; and using semi-chemicals such as pheromone attractants and trap pests.

Organic farmers have long maintained that synthetic fertilizers and pesticides increase crop susceptibility to pests[62]. Research substantiates some of these claims. Organic crops have been shown to be more tolerant as well as resistant to insect attack[63]. Organic rice is reported to have thicker cell walls and lower levels of free amino acid than conventional rice[64]. Plant susceptibility to insect herbivory has been shown in numerous studies to be associated with high plant N levels related to high inputs of soluble N fertilizers[65]. Free amino acids, associated with high N applications, have been reported to increase pest attack[66].

Soil-borne root diseases are generally less severe on organic farms than conventional farms, while there were no consistent differences in foliar diseases between the systems. The successful control of root diseases in organic systems is likely to be related to the use of long and diverse crop rotations, crop mixtures and regular application of organic amendments[67]. Increased levels of soil microbial activity leading to increased competition and antagonism in the rhizosphere, the presence of beneficial root-colonizing bacterial and increased levels of vesicular-arbuscular my-corrhizal colonization of roots have all been identified as contributing factors in the control of root diseases[68].

Organic Agriculture: Its Relevance to Indian Farming

Only 30% of India's total cultivable area is covered with fertilizers where irrigation facilities are available and in the remaining 70% of arable land, which is mainly rain-fed, negligible amount of fertilizers is being used. Farmers in these areas often use organic manure as a source of nutrients that are readily available either in their own farm or in their locality. The northeastern region of India provides considerable opportunity for organic farming due to least utilization of chemical inputs. It is estimated that 18 million hectare of such land is available in the NE, which can be exploited for organic production. With the sizable acreage under naturally organic/default organic cultivation, India has tremendous potential to grow crops organically and emerge as a major supplier of organic products in the world's organic market[69].

The report of the Task Force on Organic Farming[70] appointed by the Government of India also observed that in vast areas of the country, where limited amount of chemicals is used and have low productivity, could be exploited as potential areas for organic agriculture. Arresting the decline of soil organic matter is the most potent weapon in fighting against unabated soil degradation and imperilled sustainability of agriculture in tropical regions of India, particularly those under the influence of arid, semiarid and sub-humid climate. Application of organic manure is the only option to improve the soil organic carbon for sustenance of soil quality and future agricultural productivity[71].

It is estimated that around 700 mt of agricultural waste is available in the country every year, but most of it is not properly used. This implies a theoretical availability of 5 tonnes of organic manure/hectare arable land/year, which is equivalent to about 100 kg NPK/ha/yr[72]. However, in reality, only a fraction of this is available for actual field application. Various projections[72,73] place the tapable potential at around 30% of the total availability. There are several alternatives for supply of soil nutrients from organic sources like vermicompost, biofertilizers, etc. Technologies have been developed to produce large quantities of nutrient-rich manure/compost. There are specific biofertilizers for cereals, millets, pulses and oilseeds that offer a great scope to further reduce the gap between nutrient demand and supply. There is no doubt that organic agriculture is in many ways a preferable pattern for developing agriculture, especially in countries like India.

Conclusions

The interest in organic agriculture in developing countries is growing because it requires less financial input and places more reliance on the natural and human resources available. Studies to date seem to indicate that organic agriculture offers comparative advantage in areas with less rainfall and relatively low natural and soil fertility levels. Labour realizes a good return and this is important where paid labour is almost non-existent. Organic agriculture does not need costly investments in irrigation, energy and external inputs, but rather organic agricultural policies have the potential to improve local food security, especially in marginal areas.

Possibly, the greatest impact of organic agriculture is on the mindset of people. It uses traditional and indigenous farming knowledge, while introducing selected modern technologies to manage and enhance diversity, to incorporate biological principles and resources into farming systems, and to ecologically intensify agricultural production. Instead of being an obstacle to progress, traditions may become an integral part of it. By adopting organic agriculture, farmers are challenged to take on new knowledge and perspectives, and to innovate. This leads to an increased engagement in farming which can trigger greater opportunities for rural employment and economic upliftment. Thus through greater emphasis on use of local resources and self-reliance, conversion to organic agriculture definitely contributes to the empowerment of farmers and local communities.

The following conclusions can be drawn on important issues regarding organic farming:

1) Large-scale conversion to organic agriculture would result in food shortage with the present state of knowledge and technology, as the yield reductions of organic systems relative to conventional agriculture average 10-15%, especially in intensive farming systems. However, in traditional rain-fed agriculture, organic farming has the potential to increase the yield, since 70% of total cultivable land falls in this category. Mere 5-10% increase in farm production would definitely help achieve the targetted growth rate of 4-5% in agricultural production in the Tenth Plan period.

2) Organic manure is an alternative renewable source of nutrient supply. A large gap exists between the available potential and utilization of organic wastes. However, it is not possible to meet the nutrient requirements of crops entirely from organic sources, if 100% cultivable land is converted to organic farming.

3) Organic farming systems can deliver agronomic and environmental benefits both through structural changes and tactical management of farming

systems. The benefits of organic farming are relevant both to developed nations (environmental protection, biodiversity enhancement, reduced energy use and CO_2 emission) and to developing countries like India (sustainable resource use, increased crop yields without over-reliance on costly external inputs, environment and biodiversity protection, etc.).

4) Organic foods are proved superior in terms of health and safety, but there is no scientific evidence to prove their superiority in terms of taste and nutrition, as most of the studies are often inconclusive.

5) Combination of lower input costs and favourable price premiums can offset reduced yields and make organic farms equally and often more profitable than conventional farms. However, studies that did not include organic price premiums have given mixed results on profitability. Thus it is the premium price on the organic food which decides the economic feasibility of organic farming, at least at the current rate of development in organic agriculture.

6) In organic farming systems, pest and disease management strategies are largely preventive rather than reactive. In general, pest and disease incidence is less severe in organic farms compared to conventional farms.

(P Ramesh and A Subba Rao are in the Indian Institute of Soil Science, Nabi Bagh, Bhopal, India, can be reached at pramesh@iiss.ernet.in

Mohan Singh is in the Indian Institute of Pulses Research, Kanpur, India.)

References

1. Subba Rao, I. V., Soil and environmental pollution – A threat to sustainable agriculture. *J. Indian Soc. Soil Sci.*, 1999, 47, 611–633.

2. Pretty, J., The real costs of modern farming. *Agric. Syst.*, 2000, 65, 113–136.

3. Kortbech-Olesen, R., Export opportunities of organic food from developing countries. In *World Organics*, Agra Europa (London) Ltd, London, UK, 9–10 May 2000.

4. FAO, Food safety and quality as affected by organic farming. Agenda item 10.1. In Twenty-Second FAO Regional Conference for Europe, Porto, Portugal, Food and Agriculture Organization of the United Nations, 24–28 July 2000.

5. International Federation of Organic Agriculture Movements, First draft of 2002, IFOAM basic standards for organic production and processing. On-line report, Germany, 2001; Tholey-Tholey, *www.ifoam.org*.

6. FAO, Guidelines for the production, processing, labelling and marketing of organically produced foods. Joint FAO/WHO Food Standards Program Codex Alimentarius Commission, Rome, CAC/ GL 32, 1999, p. 49.

7. Chhonkar, P. K., Organic farming: Science and belief. Dr R. V. Tamhane Memorial Lecture delivered at the 68th Annual Convention of the Indian Society of Soil Science, CSAU&T, Kanpur, 5 November 2003.

8. Stockdale, E. *et al.*, Agronomic and environmental implications of organic farming systems. *Adv. Agron.*, 2000, 70, 261–327.

9. Stanhill, G., The comparative productivity of organic agriculture. *Agric. Ecosyst. Environ.*, 1990, 30, 1–26.

10. Wynen, E., Economics of organic farming in Australia. In *The Economics of Organic Farming* (eds Lampkin, N. H. and Padel, S.), CAB, Wallingford, UK, 1994, pp. 185–199.

11. Halberg, N. and Kristensen, I. S., Expected crop yield loss when converting to organic dairy farming in Denmark. *Biol. Agric. Hortic.*, 1997, 14, 25–41.

12. Offermann, F. and Nieberg, H., Economic performance of organic farms in Europe. *Organic Farming in Europe: Economics and Policy 5*, University of Hohenheim, Hohenheim, 1999.

13. Rajendran, T. P., Venugopalan, M. V. and Tarhalkar, P. P., Organic cotton farming in India. Central Institute of Cotton Research, Technical Bulletin No. 1/2000, Nagpur, 2000, p. 39.

14. Kler, D. S., Sarbjeet Singh and Walia, S. S., Studies on organic versus chemical farming. Extended summaries vol. 1, 2nd International Agronomy Congress, 26-30 November 2002, New Delhi, pp. 39-40.

15. Huang, S. S., Tai, S. F., Chen, T. C. and Huang, S. N., Comparison of crop production as influenced by organic and conventional farming systems. *Taichung Dist. Agric. Improvement Stn., Spec. Publ.*, 1993, 32, 109–125.

16. Singh, G. R., Chaure, N. K. and Parihar, S. S., Organic farming for sustainable agriculture. *Indian Farm.*, 2001, 52, 12–17.

17. Dormaar, J. F., Lindwall, C. W. and Kozub, G. C., Effectiveness of manure and commercial fertilizer in restoring productivity of an artificially eroded dark brown chernozemic soil under dryland conditions. *Can. J. Soil Sci.*, 1988, 68, 669–679.

18. Lockeretz, W., Shearer, G. and Kohl, D. H., Organic farming in the corn belt. *Science*, 1981, 211, 540–546.

19. Petersen, C., Drinkwater, L. and Wagoner, P., The Rodale Institute Farming System Trial: The first 15 years. The Rodale Institue, Kutztown, PA, 1999, p. 40. *www.rodaleinstitute.org*.

20. Peters, S. E., Conversion to low-input farming systems in Pennsylvania, USA: An evaluation of the Rodale Farming Systems Trial and related economic studies. In *Economics of Organic Farming* (eds Lampkin, N. H. and Padel, S.), CAB, Wallingford, UK, 1994, pp. 265–284.
21. Smolik, J. D., Dobbs, T. L. and Rickerl, D. H., The relative sustainability of alternative, conventional and reduced till farming system. *Am. J. Alternative Agric.*, 1995, 10, 25.
22. Pretty, J. and Hine, R., Reducing food poverty with sustainable agriculture: A summary of new evidence. SAFE Research Project, University of Essex, Wivenhoe Park, UK, 2001, p. 136.
23. Liebhardt, W. C., Andrews, R. W., Culik, M. N., Harwood, R. R., Janke, R. R., Radke, J. K. and Rieger-Schwartz, S. L., Crop production during conversion from conventional to low-input methods. *Agron. J.*, 1989, 81, 150–159.
24. Neera, P., Katano, M. and Hasegawa, T., Comparison of rice yield after various years of cultivation by natural farming. *Plant Prod. Sci.*, 1999, 2, 58–64.
25. Padel, S. and Lampkin, N. H., In *The Economics of Organic Farming: An International Perspective*, CAB International, Wallingford, UK, 1994.
26. Avery, D. T., Saving the planet with pesticides and plastic: The environmental triumph of high-yield farming. Hudson Institute, Indianapolis, 1995, p. 432.
27. Trewavas, A. J., The population/biodiversity paradox. Agricultural efficiency to save wilderness. *Plant Physiol.*, 2001, 125, 174–179.
28. Woodward, L., Can organic farming feed the world? *Ecol. Farm.*, 1998, 19, 20–21.
29. Lampkin, N. H., Estimating the impact of widespread conversion to organic farming on land use and physical output in the United Kindom. In *Economics of Organic Farming* (eds Lampkin, N. H. and Padel, S.), CAB, Wallingford, UK, 1994, pp. 353–359.
30. Padel, S. and Lampkin, N. H., Farm-level performance of organic farming systems: An overview. In *Economics of Organic Farming* (eds Lampkin, N. H. and Padel, S.), CAB, Wallingford, UK, 1994, pp. 201–219.
31. Olson, K. D., Langley, J. and Heady, E. O., Widespread adoption of organic farming practices: Estimated impacts on US agriculture. *J. Soil Water Conserv.*, 1982, 37, 41–45.
32. Seemueller, M., Der Einfluss unterschiedlicher Landbewirtscha-ftungssysteme ouf die Ernaehrungssitutation in Deutschland in Ab-haengigkeit des Konsumvehaltens der Verbraucher. Oeko-Institut e. V Freiburg, DE, 2000, p. 114.
33. Morabia, A., Bernstein, M. S., Heritier, S. and Beer-Borst, S., A Swiss population-based assessment of dietary habits before and after the March 1996 'mad cow disease' crisis. *Eur. J. Clin. Nutr.*, 1999, 53, 158.

34. International Federation of Organic Agriculture Movements, *IFOAM Basic Standards for Organic Production and Processing*, IFOAM Publications, Tholey-Tholey, Germany, 1998.

35. Mohan Singh, Organic farming prospects in Indian agriculture. Souvenir of 68th Annual Convention of Indian Society Soil Science, CSAU&T, Kanpur, 4–8 November 2003, pp. 52-60.

36. All-India Coordinated Research Project on Microbiological Decomposition and Recycling of Farm and City Wastes. *Progress Report*, Indian Institute of Soil Science, Bhopal, India, p. 85.

37. Veeresh, G. K., Organic farming: Ecologically sound and economically sustainable. Paper presented at the International Conference on Ecological Agriculture: Towards Sustainable Agriculture, Punjab Agricultural University, 15–17 November 1997.

38. Ramaswami, P. P., Recycling of agricultural and agro-industry wastes for sustainable agricultural production. *J. Indian Soc. Soil Sci.*, 1999, 47, 661–665.

39. Kler, D. S., Kumar, A., Chhinna, G. S., Kaur, R. and Uppal, R. S., Essentials of organic farming – A review. *Environ. Ecol.*, 2001, 19, 776–798.

40. Kler, D. S., Uppal, R. S., Kumar, A. and Kaur, G., Imbalance and maintenance of eco-system – A review. *Environ. Ecol.*, 2002, 20, 20–48.

41. Stolze, M., Piorr, A., Haring, A. and Dabbert, S., The environmental impact of organic farming in Europe. In *Organic Farming in Europe: Economics and Policy*, University of Hohenheim, Hohen-heim, Germany, 2000.

42. Conway, G. and Pretty, J. N., *Unwelcome Harvest: Agriculture and Pollution*, Earthscan Publications, London, 1991, p. 645.

43. Jeyaratnam, J., Acute pesticide poisoning: A major problem. *World Health Stat. Q.*, 1990, 43, 139–144.

44. Avery, D., The hidden dangers of organic food, Hudson Institute, 1998, *www.hudson.org*.

45. Pell, A. N., Manure and microbes: Public and animal health problem? *J. Dairy Sci.*, 1997, 80, 2673–2681.

46. Burros, M., Anti-organic and flawed. *New York Times*, 17 February 1999.

47. Jones, D. L., Potential health risks associated with the persistence of *Escherichia coli* O157 in agricultural environments. *Soil Use Manage.*, 1999, 15, 76–83.

48. Rutenberg, J., Report on organic foods challenged. *New York Times*, 31 July 2000, p. C 11.

49. Woese, K., Lange, D., Boess, C. and Bogl, K. W., A comparison of organically and conventionally grown foods – Results of a review of the relevant literature. *J. Sci. Food Agric.*, 1997, 74, 281–293.

50. Benbrook, C. and Baker, B., Placing pesticide residues and risk in perspective: Data-driven approaches for comparative analyses of organic, conventional and IPM-grown food. In *Ecological Farming Conference*, Monterey, CA, 24–27 January 2001, *www.eco-farm.org.*

51. Muramoto, J., Comparison of nitrate content in leafy vegetables from organic and conventional farms in California. *Org. Farm. Res. Found. Newsl.*, 2000, 8, *www.ofrf.org.*

52. Lampkin, N., *Organic Farming*, Farming Press, Ipswich, UK, 1990, p. 715.

53. Reganold, J. P., Glover, J. D., Andrews, P. K. and Hinman, H. R., Sustainability of three apple production systems. *Nature*, 2001, 410, 926–930.

54. Worthington, V., Nutrition and biodynamics: Evidence for the nutritional superiority of organic crops. *Biodynamics*, 1999, 224, 58–69.

55. Anon, Greener groceries at a price. Which? *Consumer Association*, UK, October 1992, pp. 28-30.

56. Benge, J. R., Banks, N. H., Tillman, R. and De Silva, H. N., Pair-wise comparison of the storage potential of kiwi fruit from organic and conventional production systems. *N. Z. J. Crop Hortic. Sci.*, 2000, 28, 147–152.

57. Warman, P. R. and Harvard, K. A., Yield, vitamin and mineral contents of organically and conventionally grown potatoes and sweetcorn. *Agric. Ecosyst. Environ.*, 1998, 68, 207–216.

58. Sellen, D., Tollamn, J. H., McLeod, D., Weersink, A. and Yiridoe, E., In The economics of organic and conventional horticulture production., Department of Agricultural Economics and Business, University of Guelph, Canada, 1993.

59. Hanson, J. C., Lichtenberg, E. and Peters, S. E., Organic versus conventional grain production in the mid-Atlantic: An economic and farming system overview. *Am. J. Alternative Agric.*, 1997, 12, 2.

60. Welsh, R., The economics of organic grain and soybean production in the Midwestern United States. Henry A. Wallace Institute for Alternative Agriculture, 1999, p. 34, *www.hawiaa.org.*

61. FAO, Organic agriculture and the environment. Food and Agricultural Organization, Rome, 2003, *www.fao.org/organicag.*

62. Yepsen, R. B., Organic plant protection: A comprehensive reference on controlling insects and diseases in the garden, orchard and yard without using chemicals. Emmaus, PA, Rodale Press, 1976, p. 688.

63. Lotter, D. W., Grannet, J. and Omer, A. D., Differences in grape phylloxera-related grapevine root damage in organically and conventionally managed vineyards in California. *Hortscience*, 1999, 34, 472–473.

64. Kajimura, T., Fujisaki, K. and Nakasuji, F., Effect of organic rice farming on leafhoppers and planthoppers. 2. Amino acid content in the rice phloem sap and survival rate of planthoppers. *Appl. En-tomol. Zool.*, 1995, 30, 17–22.

65. Phelan, P. L., Soil management history and the role of plant mineral balance as a determinant of maize susceptibility to the European corn borer. In *Entomological Research in Organic Agriculture: Selected Papers from a European Workshop*, Austrian Federal Ministry of Science and Research, Vienna, Austria, 1999, pp. 25–34.

66. Hedin, P. A., Williams, W. P., Buckley, P. M. and Davis, F. M., Arrestant responses of southwestern corn borer larvae to free amino acids: Structure–activity relationships. *J. Chem. Ecol.*, 1993, 19, 301–311.

67. Van Bruggen, A. H. C., Plant disease severity in high-input compared to reduced-input and organic farming systems. The American Phytopathological Society, St. Paul, Minnesota, 1995.

68. Azcon Aguilar, C. and Barea, J. M., Arbuscular mycorrhizas and biological control of soil-borne pathogens – An overview of the mechanism involved. *Mycorrhiza*, 1996, 6, 457–464.

69. Organic farming in India, Eximbank report, 2002; *www.Eximbankagro.com.*

70. Report of Task Force on Organic Farming, Department of Agriculture and Cooperation, Ministry of Agriculture, Government of India, 2001, p. 76.

71. Katyal, J. C., Organic matter maintenance: Mainstay of soil quality. *J. Indian Soc. Soil Sci.*, 2000, 48, 704–716.

72. Tandon, H. L. S., In *Plant Nutrient Needs, Efficiency and Policy Issues: 2000–2025*, National Academy of Agricultural Sciences, New Delhi, 1997, pp. 15–28. 73. Katyal, J. C., *Proc. Indian Natl. Sci. Acad., Part B*, 1993, 59 161.

Index